ANIMAL STORIES: TAME & WILD

ANIMAL STORIES: TAME & WILD

James Herriot
David Attenborough
James A. Michener
Desmond Morris
Philippe Cousteau
and others

Compiled by Gilbert and John Phelps

STERLING PUBLISHING CO., INC.   New York

Color art by Sheila Wright

Library of Congress Cataloging in Publication Data

Animals tame & wild.
  Animal stories, tame & wild.

  Originally published as: Animals tame & wild. 1979.
  Includes index.
  1. Zoology—Addresses, essays, lectures.   2. Pets—
Addresses, essays, lectures.   I. Herriot, James.
II. Phelps, Gilbert.   III. Phelps, John.   IV. Title.
[QL81.A54   1985]   591   85-12575
ISBN 0-8069-4722-5

Published in 1985 in the United States of America
by Sterling Publishing Co., Inc.
Two Park Avenue, New York, N.Y. 10016
First published in 1979 under the title
"Animals Tame and Wild" © 1979 by Topaz
Publishing, Ltd. and published in the
United Kingdom by Angus and Robertson
Publishers, London, Sydney, Melbourne,
Singapore, Manila
Distrubuted in Canada by Oak Tree Press Ltd.
% Canadian Manda Group, P.O. Box 920, Station U
Toronto, Ontario, Canada M8Z 5P9
Manufactured in the United States of America

Contents

Color sections follow pages 32, 64, 96 and 128

Introduction

James Herriot took as the titles for several of his hugely successful books, lines from a traditional hymn:

All things bright and beautiful
All creatures great and small
All things wise and wonderful
The Lord God made them all.

We like the quotation because it emphasizes the fact that we all have a common ancestry.

Yet for whatever reason—the pressures of an increasingly artificial daily existence, a (often unjustified) sense of superiority, or just ignorance—many people are losing touch with the animal world.

This shows itself most obviously in the exploitation of animals for food or sport, or in an over-sentimental and fundamentally patronizing approach. Both attitudes, of course, have a common source—a lack of real understanding.

To get a truly balanced view of the animal world requires a very special kind of sensitivity which, frankly, only years of observation and shared experiences can bring.

This book aims to bring together the views and experiences of many of the world's top animal writers and naturalists and to highlight some of their most fascinating insights and experiences.

What comes through in all their writings is respect for the animal kingdom, earned not least from the realization that there are precious few of man's inventions or recently developed faculties that an animal somewhere had not already evolved, at least in their barest essentials—be it the complex building techniques of the bee or the blackcap's amazing navigational system.

We hope this book will surprise and delight the reader.

GILBERT AND JOHN PHELPS

About the Editors

Gilbert and John Phelps are father and son. Gilbert Phelps was born in Gloucester (an ancient cathedral city in the West Midlands of England) and was educated there before going on to Cambridge to study English literature. John Phelps was born in Lisbon, Portugal (where his father was working for the British Council). But John, too, has strong Gloucester ties because, like his father, he is an hereditary chartered freeman of the city, descended from a long line of yeoman farmers in and around the city.

Gilbert Phelps once worked for the BBC but left to devote himself full time to writing. He has published many books, including nine novels, and now lives in the heart of the Oxfordshire countryside with his writer wife, a son still in school, and three neurotic cats.

John Phelps, who was educated at Highgate School in London, has spent most of his working life in journalism. He has worked both in Fleet Street and in the Provinces and is now with the *Cambridge News*. He has written a book for children and he shares with his father a keen interest in sport, cinema, and, of course, animals.

UNUSUAL PETS

A Skunk in the Family

Constance Colby

It may seem rather conventional to start off a book about animals with a chapter on pets. But it is, after all, through our pets that most of us obtain our first knowledge, affection and respect for the creatures with whom we share that travelling space-ark, the earth. And there's nothing conventional about some of the animals which have been adopted as pets.

A skunk is a wild animal, even though he may have been born in captivity, and he never fully accepts domesticity. Oh, he likes people well enough—may even grow attached to one or two of them—and he certainly likes the comforts of life which they can provide. But he never comes to depend on people, as some other animal might. And, unlike a domestic animal, he never tries to please anyone but himself. It is not a skunk's nature to study your face speculatively, kitten fashion, trying to decide what you are thinking. Nor will he come beaming up to you with the typical puppy present: a lovely dead frog perhaps, or your best scarf chewed to shreds.

As for expecting a skunk to regulate his sleeping, eating or excreting habits in order to conform to some human requirement, this is totally out of the question. He is aware of human displeasure—he hates being shouted at and will try to avoid it whenever con-

venient—but it never enters his head to alter his own behavior even a fraction in order to win human approval. He simply can't be bothered.

No matter what the issue—diet, house-training, daily routine, or whatever—it is the humans who must finally adapt. Life would have been much easier for us all if I had been able to accept the inevitable more gracefully.

Skunks have an inborn sense of neatness and order. Their fur is always immaculate, their claws are polished, their dens are equally neat and clean. A mother will conscientiously seek out a new nest for each litter (with the result that internal parasites are almost never found in the young).

In the woods, skunks go about their rounds as if on schedule, appearing at the same place at almost the same time every night. They seem to have a sort of mental cupboard inventory of just what belongs where and freeze with instant suspicion if they come upon a newly fallen branch, an unfamiliar scatter of leaves, or a rock lying across a formerly clear path. Skunks are firm believers in the old-fashioned house-keeping dictum, "A place for everything and everything in its place."

They are equally precise when it comes to their toilet. The usual procedure is to back into a discreet corner, then scratch a sanitary covering of leaves over the spot, and finally scoot along the ground using the grass and moss as toilet paper.

House-training such an animal should be easy, since his own natural inclinations are so closely in accord with human requirements.

"A skunk is a very clean-living animal," our veterinarian informed us.

The pet-shop owner concurred. "Your average skunk," he assured us, "practically house-trains himself."

Which may be true of *his* average skunk.

Certainly the recommended system could not be simpler. In any new environment, a skunk will quickly select a preferred spot—usually a corner—for defecation. Once he has established his preference, you cover the area with newspapers or set up a cat-litter pan there. After he has grown used to this, you have only to transfer the papers or pan to whatever place you yourself have selected and the skunk will follow along.

Unless, of course, his liking for his chosen corner is stronger than for newspapers or cat litter. A good corner, after all, has much to recommend it. It is quiet and private. Also, nothing can creep up on you from behind. A patch of newspapers, on the other hand, or a pan of unnatural-smelling sand, offers neither privacy nor security—especially if it has an alarming habit of suddenly vanishing and then reappearing in a totally new spot.

It is really not so surprising that a baby skunk might regard house-training paraphernalia with suspicion. At least, ours did.

True to expectation, he picked out a favorite corner almost at once. Unfortunately, this was exactly in the angle between two of the most important doors in the back kitchen. By day this was the junction of the most heavily trafficked routes in the house, which could prove to be a problem later on if the little skunk gave up his nocturnal schedule. In addition, the area was too small to accommodate his beautiful new turquoise cat pan.

Actually, none of the corners in the back kitchen was what might be called the ideal place, from the human point of view, but the corner near the birdseed cupboard (which was now secretly wired shut) would be the least inconvenient. All we had to do was persuade him to transfer from one corner to the other.

By now it was clear that building barricades was not the answer. But a skunk nose is highly sensitive. Perhaps smell would be a training aid.

Hoping to attract him to the approved corner, we covered the floor with already used papers. There was no question but that they did smell. However the lure was ignored.

Then we tried to make the other corner as unattractive as possible. We sprayed it with "Doggy-Do-Not!" and "Off!" and every insect repellent. The back kitchen not only looked like a town dump, it was beginning to smell like one.

Strong odors, however, are no deterrent to a skunk, as we perhaps should have known in advance. Unperturbed, he continued to use his favorite corner between the doors.

Then Jim had a stroke of pure inspiration. He opened both of the doors, back to back, and fixed them in this position by looping an old jump rope around the knobs. And presto!—*the corner was gone.*

This brilliant stratagem worked. When his special nook disappeared, the little skunk had no choice but to retreat to the other corner. Moreover, after a few days he seemed to grow resigned to the new location —which was fortunate because a streak of rainy weather made it imperative to close the outside door.

When the original corner reappeared, we piled it high with hefty rocks, just in case. They would not stop him from using the area if he were determined to; they would, however, make it inconvenient. But although the rocks were shoved around a bit every night, they did not seem to be needed. The habit that drew him to the new place had finally become established. Eventually, the rocks were discarded.

For a long time afterwards, however, whenever he was angry he defiantly returned to his old spot in the long-prohibited corner to leave a token of his displeasure.

If potty training the little skunk was a problem, naming him was not so easy either. The scientific designation turned out to be *Mephitis mephitis* (*mephitis* meaning "noxious vapor" doubled!) and the Algonovian Indian name, from which the word "skunk" is derived, was *seganku* (we've no idea what that one means). We didn't feel that we could cope with either of these.

In the end, the little skunk's own nature suggested the right name for him, and when we tried it out, he recognized it at once.

It all happened because a thirteen-year-old neighbor came over one day to take some animal pictures.

He arrived draped with equipment and immediately began surveying the house for all the best picture locations. He tilted lamps and rearranged furniture and pinned up a sheet to serve as a backdrop for close-ups. When everything was ready, he turned to me and said, "All right, where is he?"

Where indeed? Ordinarily at that hour a sensible house skunk would still be sleeping. But we had made a point of waking ours in mid afternoon to be sure that he would be ready to have his picture taken. Alice had given him several nice pieces of bacon and then Anne had raced him around the house to put him in a playful mood. However, in the excitement of Davy's arrival, we had forgotten him just long enough.

We searched, as the saying goes, high and low—mainly low. We crawled around peering under every bed and chair and chest. The skunk had simply vanished.

"It's not that he's unfriendly," Anne explained to Davy. "He's not hiding because he doesn't want to meet you. It's just that he likes to be secret."

"Then what's he doing chewing up my Nikon strap?" Davy asked. Sure enough there he was, cautiously sampling the leather strap which was dangling over the edge of the armchair.

We never found the secret hiding place he used that day. But we had found a name for him.

Ruler of secret trails and hidden nests; inventor of private games which kept him racketing around half the night; planner of surreptitious assaults on garbage pail and refrigerator—this small animal cherished secrecy above all else. And when, for the first time, we called, "Here, Secret!" he came at once.

Months later a visitor congratulated us on choosing such an appropriate name for a skunk. "Just like the deodorant, how clever!" We have never heard of the product, but we soon discovered that lots of people had. By then, however, it was too late to change. So Secret he remained, or sometimes Seekie, which wasn't too bad either, since seeking (mostly for food) was one of his favorite pastimes.

And in case you've been wondering about the notorious smell of the skunk—*that* can be dealt with.

For a wild skunk, of course, the ability to produce a powerful odor—and with it, a burning spray—is all-important. Often his life may depend on it.

This unusual method of self-defense is the result of a highly specialized development of the anal glands. Dogs and many other animals have similar glands but none so well developed.

The skunk's two scent glands are embedded in the muscle tissue on either side of the rectum. They produce an amber-colored fluid (whose main element, butyl mercaptan, was formerly used in making certain perfumes to add a clinging quality) which can be sprayed out through retractable ducts. The ends of these ducts shoot out through the sides of the anus and then retract again with lightning speed, thus preventing any of the fluid from getting on the skunk himself; it is the victim that reeks, not the skunk. (In the descenting operation these two glands must be removed with great care so as not to damage the rectal wall. The task is especially difficult—for all concerned—because most veterinarians prefer not to use an anaesthetic, as it is difficult to estimate the correct amount.)

The skunk may fire off one gland or both together in a powerful spray, and he can repeat the process again and again if necessary. His aim is good for ten feet or more; the odor has been known to carry at least half a mile. The fluid stings the skin and burns the eyes, sometimes causing temporary blindness. The smell alone makes most animals sick. What is more, it does not begin to fade until days later. An encounter with a skunk is not easily forgotten.

From: *A Skunk in the Family*, by Constance Colby.

A Sloth in the Family

Herman Tirler

Or what about a sloth as a pet? The Tirler family adopted one (called Nepomuk) when they were living in Brazil—where the animal is known as the Ai, because the only sound he ever makes is a very occasional deep sigh that sounds like "a . . . i."

There is a tree in our garden which hummingbirds, called "flower-kissers" in Brazil, delight in. One day, when Nepomuk was up in this tree, a hummingbird flew close and then hung there in the air with thrumming wings, observing the sloth. Suddenly there was a loud crack among the branches, the hummingbird darted away and poor Nepomuk tumbled out of the tree and landed with a thud on the ground, where he sat gazing round him with a look of amazement on his face!

What had happened was this: when climbing, a sloth often gropes blindly; his claws seize on to anything within reach in a strong and lightning reflex action, regardless of what the object may be. On this occasion it was his own tail, so no wonder he lost his balance.

In spite of such accidents it is usually a pleasant sight to watch a sloth climbing, for its slow, deliberate movements are very graceful. On the ground, the poor ai is a very different case, for then he becomes a helpless creature unable even to raise himself. To move, he has to stretch out his forelegs and pull himself along labouriously, digging his claws into the ground. His short hind legs do what they can to help, but that is not much and the sloth's belly never clears the ground. In sandy soil he leaves a most peculiar track. On the polished floors in our house, Nepomuk is prone and helpless.

Is it true, as the old book I read said, that he "doth squander twain whole days" to get from one tree to another? Certainly not, because sloths seldom come down to the ground to cross from one tree to another.

When a storm goes roaring through the forest, it is Nepomuk's delight to take to the tree-tops, where he gives a display of breathtaking acrobatics. Climbing to the topmost branches, he will let his weight pull him into a horizontal position, pointing towards the neighbouring tree; then, cautiously distributing his weight between a number of twigs, he will edge his way to the very end, and, while the rain pours off him and the wind jerks him this way and that, he will make a long arm at the next tree-top as it swings towards him and away again; then, as it returns on the next gust, things happen in a rush: Nepomuk catches hold of a twig of the swaying crest, then swiftly seizes another two; but often they are obviously far too weak to climb on to and already the tree-top is swinging back. At this point, if I am watching the proceedings, my heart leaps into my mouth as I see Nepomuk crash into the foliage of the next tree. But I need not worry, for there he hangs, unperturbed, swaying to and fro until he finally comes to rest.

From: *A Sloth in the Family*, by Herman Tirler.

Bears in the Family

Peter Krott

Yes, they have become family pets too—well, more or less. Peter Krott had two Alpine bear cubs named Bumsli and Sepha.

At long last spring had come to the bear country, and the snow was rapidly disappearing. Bumsli and Sepha fed on crocuses, just like the deer and mountain hares. They nibbled willow and alder catkins, but carefully avoided the woods where there was still snow about. Hibernation had come to an end.

The brown bear has always enjoyed great popularity. Very early he was made the object of cults in many parts of the world. He is a favourite in zoos and circuses, and as a model for toys and mascots. This, I believe, is due to the fact that, though a big and strong animal, the bear looks friendly, and at times moves in an almost human way. We do not always realise sufficiently how much our attitude towards living things depends on what they look like and how they move about. Closer acquaintance with the bear will soon show that he is not quite so easy to deal with. Like all animals who have no natural enemies other than hunger and disease, the bear cannot be domesticated. There can never be a friendly relationship between man and bear. We succeeded in getting on well with Bumsli and Sepha because they took my wife and me as their kind, and considered our children as their brothers. We had to keep strictly to the rules of the game and our roles as parents if we wanted to avoid accidents. The simple truth was that our bears were not tame, but that we had become bear-like.

It was impossible for outsiders to have the same understanding of the psychology of bears, or for us to ask them to be bear-like. We therefore thought it best to move with our bears to a region with fewer chances of encountering strangers, where it would then be easier to complete our experiment. I spent some sleepless nights wondering where to go, as there were few suitable places left.

At Easter my wife and the children came to the Alps, and Bumsli and Sepha knew them at once. They sucked my wife's hands, and disappeared with the children into the little shed where I had kindled a good fire. Both bears liked to watch the leaping flames, sometimes aiming blows at them with their paws. They never burnt themselves, and did not appear frightened. The Alpine bear is not afraid of the fires made by the herdsmen.

It was a little uncanny to me to see my children with the bears, who were now so much bigger than Martin and Max. I settled in a deck-chair in front of the hut, while all my family did their best to disturb my peace. Suddenly Sepha wanted to play with me. Nimbly she jumped from the roof of the little shed straight into my lap, tearing my newspaper at the most thrilling story. The deck-chair burst, but this did not prevent Bumsli from trying to lie down on it. In the midst of this to-do, two strangers—forestry workers—turned up and stared at us open-mouthed. The bears made threatening noises, and blowing angrily sat up on their hind legs, next to the boys who had come out of the shed. The men were quite glad to comply with my request to go away!

In the afternoon we all took our usual Sunday walk. We noticed that Bumsli and Sepha were more nervous and more jealous than they had been the previous summer. I went ahead with the boys while the bears lagged behind, eating crocuses. My wife closed the hut, and followed us. The bears scarcely had time to notice her sitting on a rock with the boys, when they rushed up to me, overcome with jealousy. They were no longer interested in the crocuses. They dashed off again when she came up with a laugh, and only dared join us when she called them.

We all enjoyed the little patches of snow. The boys and bears slid down them, and the boys vigourously engaged in a hectic snowball fight. Bumsli rolled happily on his back, made a snowball with his paws, and played with it. Sepha, too, loved rolling in the snow, pushing her paws in the air—it was great fun! Suddenly a caper-kailzie fluttered to the wood's edge, and a jay gave his warning call. We all looked for cover against the new intruders.

There they were: two tourists laden with sleeping-bags and cameras, standing at the top of the meadow, inquiring for sleeping quarters, and making ready to take photographs. Surely they could read, and must have seen my warning notices? I told them to be off as quickly as possible. My angry voice affected the bears, and when the two enthusiastic photographers fled downhill, Bumsli and Sepha set out in hot pursuit, which made the two run very quickly indeed.

From: *Bears in the Family*, by Peter Krott.

A Hare about the House

Cecil S. Webb

Bears can be big trouble, but as Cecil Webb recounts in this story, the combination of a hare named Horrie and a rabbit called Squirt can cause the fur to fly too!

When Horrie was only sixteen days old there was an incident which led to a further addition to the family. One morning the lad looking after the Zoo's stock of domestic rabbits arrived at the house carrying one of a litter that he had rescued. The mother, a large white rabbit, had eaten the rest of her offspring and was about to devour this one. In spite of its appearance my wife took pity on it and took it in. It was naked, blind and ugly—looking more like a tiny embryo of a bull terrier than a young rabbit. To make it more

comical it walked like a dog instead of hopping. By comparison with the beautiful, wide-eyed, alert, fluffy little leveret, this was one of nature's monstrosities. However, it was interesting to see the difference between the young of the two rodents of somewhat similar size, the one born above ground—perfectly developed—and the other normally born in a burrow—blind and helpless—and in our kitchen bumping into everything as it walked along.

When Horrie's donor arrived for tea one day, she was quite aghast when she saw my wife hand-feeding the new arrival. She could only exclaim, "Oh no, how could you?"—but my wife had different ideas; ugly or not ugly, its life had to be saved, and after all, the "ugly duckling" might develop into a "beautiful swan."

For a while he was called Harvey, but as I greeted him so often with the remark, "poor little squirt," his name henceforth became Squirt.

At first Horrie regarded this strange object with complete indifference, and even when it staggered blindly quite close to him on the kitchen floor, he entirely ignored it.

It is in Squirt's favour that even while still blind he never once wet his bed but always waited to be lifted out of his box into the sand-tray, and also made gallant attempts to wash himself. He quickly grew fur and after twelve days his eyes opened and from then on, Horrie, although very much a babe himself, took on the role of guardian to his comparatively backward companion, washing him from head to foot and apparently worried if Squirt was not looking as immaculate as himself. The attachment grew and grew and they became inseparable companions to a degree rarely seen in animals of any kind,

Horrie was in fact as clean as he was playful:

The toilet of animals is an important business and many have special adaptations for this. For instance, the lemurs have their lower incisors and canines grouped together like a comb and they project horizontally. Their sole function is for toilet purposes. A further toilet arrangement is the development of a claw on the second toe of each foot, in place of the flat nails on all the other toes and fingers.

Judging by the time Horrie spends in doing his own toilet, hares must be about the most fastidious of all animals. He works systematically, using his feet, tongue and teeth. As a rule he commences with his ears, first pulling down one with a front paw. This remains so while he gently combs downwards with his thumb or toilet claw which is placed high up on the foot. The whole action reminds one irresistibly of a woman combing her long tresses. With his head held low he holds the ear down to touch the ground and then licks the inner surface near the tip. When face-washing he sits up like a squirrel, licking each front paw in turn and rubbing it over his head. His coat is groomed by the combined action of licking with his tongue and combing with his teeth, special attention being paid to any part where the hairs are bunched together or knotted, and then a great deal of tugging takes place. To get to the back of his neck he uses one of his long hind legs after first licking the paw. The disproportionate length of the hind leg of the hare makes this action look decidedly comical, but even more so is his effort to clean inside his ears with a back paw.

Some of Horrie's toilet attitudes are unbelievably graceful, reminding one at times of a ballerina and at others of a fawn. Some again are quite like those of a contortionist, and not the smallest portion of his coat is missed. Even the hairy soles of his feet come in for more tugging to get rid of any foreign matter. To clean his hind feet, one leg at a time is thrust forward till it sticks well in front of his head; he then bends forward while spreading the foot like a great hand, and cleans between the toes. The knees are attended to by holding a hind leg vertically with the whole of his foot well above his head in a most ludicrous fashion. (Squirt washes as little as possible and leaves his toilet mainly to Horrie.)

The early morning is playtime for Horrie and he loves a romp before going to sleep, which he does for a good part of the day. His playtime act is a sight for the gods, especially when he is in a skittish mood. The dining-room is large (twenty-eight by nineteen feet), and has a thick red carpet which means that Horrie can get up speed and corner abruptly without skidding. To get him really excited it is best to chase him clapping one's hands, and then he tears off leaping and twisting in the air and zig-zagging at great speed. Sometimes I have seen him go flat out round and round the long table like a greyhound on a track, until I became quite dizzy watching him. Much as he likes the dining room to skip and race in, he is, perhaps, happiest when he is free to roam the house. This gives him real scope for his playfulness and obviously gives him immense satisfaction, for when it is over he is no longer restless—he asks for nothing more than sleep. He loves tearing up and down stairs, going up two at a time with a most beautiful action. Once on the upper landing the urge to dance is irresistible. In fact, one might be excused for saying that he becomes a trifle mad. A stranger would definitely think so if he suddenly saw a hare streak out of a bedroom, leap into the air, disappear into another room, spring on and off all the beds, shoot out again and corner so fast that he rolled over on his back, then tear down the stairs and skid along the polished floor in the hall, kicking rugs in all directions.

Later on Horrie had some interesting encounters with the opposite sex:

Hares and rabbits have never been crossed and it is probably impossible to do so. However, out of curiosity, we decided to carry out the experiment and a handsome female was selected for him. Horrie received the damsel with his usual air of detachment, and reminded me of a camel looking across a desert.

In the meantime Squirt was eying the intruder like a buffalo, weighing up the situation before charging. He had never before seen a rabbit, so who could tell what was on his mind? We had not long to wait; he shot at her in a blind fury charging all round the pen in hot pursuit. Horrie sat serenely like a sphinx, with his nose twitching while the chase went on—he would never be guilty of such undignified behaviour himself. To save the poor female, I had to remove Squirt, leaving her with Horrie, and put Squirt in the yard where he could see the other two through the railings. As soon as he found that he was frustrated, his manner changed to one of undiluted charm. He ran and skipped along the fence, backwards and forwards, inducing the female to join in the game on the opposite side. After a while they were having such a gay game together that I thought it safe to let Squirt back into the pen. The lady had now become familiar with her surroundings and was no longer the timid creature she was when introduced. Without any preliminaries Squirt took a headlong rush at her, but this time she changed her technique. Skipping lightly over his head, she gave a downward kick with her back legs and, as she did so, caused a lump of fur to fly off his back. Squirt now had his blood up and was determined to finish the vixen off there and then, but each time he cornered her she sprang over him and gave him a pounding. There was so much white fur flying in the pen at one time that it was like watching the fight through a miniature snowstorm. This time it was Squirt who had to be rescued and to save him from further punishment we sent her back to her zoo quarters whence she came. Squirt was very much battle-scarred but at least the potential rival had gone and he returned to Horrie with head bloody but unbowed, and flung himself down at his feet as if to show that he fought for his affection.

From: *A Hare about the House*, by Cecil S. Webb.

Otter Nonsense

Gavin Maxwell

Otters are enchanting creatures, and have often been adopted as pets. In his classic, Ring of Bright Water, *Gavin Maxwell wrote about his two otters. The first of them was named Mijbil—an Arabic name because his owner had brought him to his home on the coast of the Scottish Highlands from the marshes of Southern Iraq.*

The time of getting to know a wild animal on terms, as it were, of mutual esteem, was wholly fascinating to me, and our long daily walks by stream and hedgerow, moorland and loch, were a source of perpetual delight. Though it remained difficult to lure him from some enticing stretch of open water, he was otherwise no more trouble than a dog, and infinitely more interesting to watch. His hunting powers were still undeveloped, but he would sometimes corner an eel in the mill dams, and in the streams he would catch frogs, which he skinned with a dexterity seemingly born of long practice.

Even in the open countryside he retained his passion for playthings, and would carry with him for miles some object that had caught his fancy, a fallen rhododendron blossom, an empty twelve-bore cartridge case, a fir-cone, or, on one occasion, a woman's comb with an artificial brilliant set in the bar; this he discovered at the side of the drive as we set off one morning, and carried it for three hours, laying it down on the bank when he took to water and returning for it as soon as he emerged.

In the traces left by wild otters he took not the slightest interest. Following daily the routes for which Miji expressed preference, I found myself almost imperceptibly led by his instinct into the world in which the otters of my own countryside lived, a watery world of deep-cut streams between high, rooty banks where the leaves of the undergrowth met overhead; of unguessed alleys and tunnels in reedbeds by a loch's edge; of mossy culverts and marsh-marigolds; of islands tangled with fallen trees among whose roots were earthy excavations and a whisper of the wind in the willows. As one may hear or read a strange, unusual name, and thereafter be haunted by its constant coincidental recurrence, so, now that I had through Mijbil become conscious of otters, I saw all around me the signs of their presence where I had been oblivious to them before: a smoothed bank of steep mud which they had used for tobogganing; a hollowed-out rotten tree-stump whose interior had been formed into a dry sleeping place; the prints of a broad, capable, webbed foot; a small tarry dropping, composed mainly of eel bones, deposited upon a stone in midstream. In these last I had expected Miji to show at least an equal interest to that which he had displayed in their canine counterparts, but whether because otters do not use their excreta in an anecdotal or informative way, or because he did not recognize in these the products of his own kind, he treated them as if they did not exist.

During all the time that I had left him he killed, so far as I know, only one warm-blooded animal, and then he did not eat it, for he seemed to have a horror of blood and of the flesh of warm-blooded animals. On this occasion he was swimming in a reedy loch when he caught a moor hen chick a few days old, a little black gollywog of a creature. He had a habit of tucking his treasures under one arm when he was swimming—for an otter swimming underwater uses its forelimbs very little—and here he placed the chick while he went on in a leisurely way with his underwater exploration. It must have drowned during the first minute or so, and when at length he brought it ashore for a more thorough investigation he appeared disappointed and irritated by this unwarrantable fragility; he nuzzled it and pushed it about with his paws and chittered at it in a pettish sort of way, and then, convinced of its now permanent inertia, he left it where it lay and went in search of something more co-operative.

Later, there was another otter called Edal:

Routine is, as I have explained, of tremendous importance to animals, and as soon as we saw that Edal was settled we arranged a daily sequence that would bolster her growing security. She had her breakfast of live eels, sent, as they had been for Miji, from London, and then one or other of us took her for a two-hour walk along the shore or over the hills. During these walks she would remain far closer at hand than Miji had done, and we carried the lead not so much as a possible restraint upon her as a safeguard against attack by one of the shepherd's dogs; for

Edal loved dogs, regarded them as potential playmates, and was quite unaware that many dogs in the Western Highlands are both encouraged and taught to kill otters.

On one of these morning outings with her I had a closer view of a wild otter than ever before. Edal was hunting rock-pool life on a ledge two or three yards from the sea's edge and a few feet above it; she had loitered long there among the small green crabs, butterfish and shrimps, and my attention had wandered from her to an eagle coasting over the cliffs above me. When I turned back to the sea I saw Edal, as I thought, porpoising slowly along in the gentle waves just beyond the pool where she had been. I could have touched her with, say, the end of a salmon rod. I whistled to her and began to turn away, but as I did so, perceived something unfamiliar in her aspect; I looked back, and there was a wild otter staring at me with interest and surprise. I glanced down to the pool at my feet, and saw Edal, out of sight of the sea, still groping among the weeds and under the flat stones. The wild otter stayed for a longer look, and then, apparently without alarm, resumed his leisurely progress southward along the edge of the rocks.

In those rock pools along the shore Edal learned to catch gobies and butterfish; occasionally she would corner a fullgrown eel in the hill streams, and little by little she discovered the speed and predatory powers of her race. Her staple diet was of eels sent alive from London, for probably no otter can remain entirely healthy without eels, but she was also fond of ginger nuts, bacon fat, butter, and other whimsical *hors d'oeuvres* to which her upbringing by humans had conditioned her. Among local fish she disdained the saith or coal fish, tolerated lythe and trout, and would gorge herself gluttonously upon mackerel. We put her eels alive into her pool, where after early failures in the cloud of mud that her antics stirred up, she proved able to detect and capture them even in the midst of that dense smoke-screen. This is achieved, I think, by the hypersensitive tactile perception of her hands, for when in the shallow end of the pool she would appear deliberately to avert her gaze while feeling round her in the opaque water; the palms, too, are endowed with a "non-slip" surface, composed of a number of round excrescences like the balls of fingers, which enable her to catch and hold between them an eel that would slither easily through any human grasp.

By the end of June she was swimming as an otter should, diving deep to explore dim rock ledges at the edge of the sea tangle, remaining for as much as two minutes under water, so that often only a thin track of bubbles from the imprisoned air in her fur gave guide as to her whereabouts. (This trail of bubbles, I have noticed, appears about six feet behind an otter swimming a fathom or so under water at normal speed; never, as the eye subconsciously expects, directly above the animal.) But though she lost her fear of depth she never felt secure in great spaces of water; she liked to see on at least one side of her the limits of the element as she swam, and when beyond this visual contact she was seized with a *horror vacui*, panicking into an in-

fantile and frenzied dogpaddle as she raced for land.

Hence our first experiments with her in the rowing boat were not a success; the boat was to her clearly no substitute for *terra firma*, and in it, on deep water, she felt as insecure as if she were herself overboard—more so, in fact, for she would brave a wild rush for the shore rather than remain with us in so obvious a peril.

From: *Ring of Bright Water,* by Gavin Maxwell.

A Gaggle of Odd Pets

Jeanette Travers

One unlikely pet at a time, one would have thought, would be more than enough to take on—but not for Jeanette Travers. On their island home on the River Thames she and her husband Tony introduced two leopards and four ocelots! One of the leopards is named Poppet—and her behavior can be disconcerting at times.

Gradually, I gained Poppet's confidence and this was initiated by giving her food. At least once a day, I fed her by hand so that she became used to my scent and during this period I did not wear perfume as I thought this might confuse her. I sat with her often, talking to her, but somehow I instinctively felt it would be wrong to muss her or to have the same rough and tumble I had had with Snoopy (an ocelot).

I have found that I get on better with male animals than with females. I can relax with them and they with me in a way which I never can with female animals, with the exception of female Willy, the domestic cat, but then she is neutered. People have sometimes asked me whether animals can tell the difference between male and female humans and I have no doubts that the cats can. If a female cat forms an attachment to Tony, I am always wary and treat her with respect because I know that in her eyes, I am the opposing female and that she might harm me, given the opportunity. I have tried to counteract this by always being the one to feed the females, but I am still ultra-cautious—female jealousy over a male is very strong, whether animal or human. In later years, when Poppet became adult, she would be extremely jealous of me, as the usurping female with Tony's affections, but for the moment she was a little cub which I could pick up occasionally.

It soon became an established routine at night for Poppet to use her toilet tray, jump on to our bed, romp around for a few minutes and then, with a gigantic sigh, flop down between us fast asleep. Normally, I am the first to rise in the mornings and when Poppet realized this, she would place her paw on my face at about half past six and nuzzle me until I got out of bed. She never touched Tony at this time.

One problem which I had with Poppet was that she would insist on jumping up at me when I was carrying full dishes and cups. This happened several times as I was carrying the breakfast from the kitchen, resulting in smashed crockery and porridge all over the floor. I soon learnt that just as Poppet was springing towards me, to show her that I was carrying full dishes of porridge and to say firmly, "Don't jump up!" Poppet, in turn, soon began, when she saw the porridge, to skid to a halt and then jump away without touching the dishes. I like to think that she was trying to save me mopping up yet more mess, but perhaps it really was that she just did not like my cooking!

One day, Poppet jumped on top of a glass-fronted cabinet full of my mother's valuable antique china. Her weight was too much for it and Tony and I watched with horror as the cabinet tottered and then fell slowly forward wedging halfway against a settee, breaking one of the glass panels. Tony dashed forward to prevent the cabinet from falling any further. The china was in a jumbled heap inside and I took it out as he supported the cabinet but as I was doing this, the valuable china cascaded to the ground and smashed. Poppet had, by this time, wisely disappeared under the settee. Among the smashed china was my mother's best tea service, tea cups with broken handles; treasured pieces from her girlhood days lay in fragments. It was then that we realised the value of having a tube of glue in the house and we sat on the floor until three o'clock in the morning trying to stick handles back on those cups which remained unbroken.

Poppet's attachment to Tony grew as she became older, and his to her, but I thought that she and I were good friends. However, one night when I was in bed, Poppet sat on my head and urinated over me. My hair was dripping, the urine running down my face, much to Tony's amusement. Fortunately, the urine of these jungle cats is not offensive and smells pleasantly woody, unlike that of the domestic cat. I am still not sure whether her action was that of like or dislike. She has never urinated on Tony's head though she has had ample opportunity.

The ocelots also achieved their share of mischief:

When Tony has a headache, he obtains relief if I pull his hair hard. Brutus must have watched me do this because one day, when Tony was sitting in a chair,

Brutus jumped up behind him, put his front paws on to his shoulders, took a mouthful of hair and tugged. As Tony did not chase him away, he continued until he had pulled at every hair, dribbled while he was doing it, so that Tony's hair and scalp were damp with saliva. Inexplicably, Brutus enjoys pulling human hair and it is sometimes difficult to remove him from Tony's shoulders. One day, some friends called and while the man was talking to Tony, his wife sat on the sofa with me. After a few minutes, I went into the kitchen to make some coffee, leaving her reading a magazine. Suddenly there was an urgent shriek from the girl and the next moment, Brutus came bounding into the kitchen with what looked like a lump of hair in his mouth. He was closely followed by the girl crying in an unusually high-pitched voice, "He's got my wig!"

Apparently, Brutus had jumped onto the back of the sofa and had started to pull at her hair, as usual, which came away in his mouth. I quickly snatched a piece of meat to try to distract him so that I could rescue the wig. But he raced out of the kitchen, skillfully avoiding our outstretched hands. He tripped over the wig, part of which had fallen out of his mouth and was trailing on the ground, which gave me time to catch up with him. He then began shaking and tossing the wig up in the air, but always catching it before me to streak away again out of reach. Tony returned and between us we managed to retrieve the tangled mess of hair, which was once a wig, by giving him a pink feather duster to play with instead. He was shut away in another room while we placated the girl and I made arrangements to buy her a replacement wig. When I looked in on Brutus, some time later, he was sleeping peacefully on a cushion, exhausted by his riotous morning and surrounded by pink feathers.

From: *Starting from Scratch*, by Jeanette Travers.

Mr. Worley and His Pigs

James Herriot

Veterinary surgeons don't only attend pets, of course, but a wide range of animals in all kinds of different situations—in kennels and racing stables, on the farm, in zoos and nature reserves, and so on. Their approach has to be professional and unsentimental, but they are hardly likely to have gone into the profession unless they had a genuine liking for animals—and they have unrivalled opportunities of getting to know them and of observing their relationship with other humans. In his famous book, All Creatures Great and Small, *for example, James Herriot tells us about the man who lived for his pigs:*

"I can see you like pigs," said Mr. Worley as I edged my way into the pen.

"You can?"

"Oh yes, I can always tell. As soon as you went in there nice and quiet and scratched Queenie's back and spoke to her I said, 'There's a young man as likes pigs.' "

"Oh good. Well, as a matter of fact you're absolutely right. I do like pigs." I had, in truth, been creeping very cautiously past Queenie, wondering just how she was going to react. She was a huge animal and sows with litters can be very hostile to strangers. When I had come into the building she had got up from where she was suckling her piglets and eyed me with a non-committal grunt, reminding me of the number of times I had left a pig pen a lot quicker than I had gone in. A big, barking, gaping-mouthed sow has always been able to make me move very smartly.

Now that I was right inside the narrow pen, Queenie seemed to have accepted me. She grunted again, but peaceably, then carefully collapsed on the straw and exposed her udder to the eager little mouths. When she was in this position I was able to examine her foot.

"Aye, that's the one," Mr. Worley said anxiously. "She could hardly hobble when she got up this morning."

There didn't seem to be much wrong. A flap of the horn of one claw was a bit overgrown and was rubbing on the sensitive sole, but we didn't usually get called out for little things like that. I cut away the overgrown part and dressed the sore place with our multi-purpose ointment, *ung pini sedativum,* while all the time Mr. Worley knelt by Queenie's head and patted her and sort of crooned into her ear. I couldn't make out the words he used—maybe it was pig language because the sow really seemed to be answering him with little soft grunts. Anyway, it worked better than an anaesthetic and everybody was happy, including the long row of piglets working busily at the double line of teats.

Well, it was the beginning of a continuous relationship and Mr. Worley and James Herriot would converse quite frequently on the merits of the porcine species:

On one occasion, in the middle of a particularly profound discussion on the ventilation of farrowing houses, Mr. Worley stopped suddenly and, blinking rapidly behind his thick spectacles, burst out:

"You know, Mr. Herriot, sitting here talking like this with you, I'm, 'appy as the king of England!"

His devotion resulted in my being called out frequently for very trivial things and I swore freely under my breath when I heard his voice on the other end of the line at one o'clock one morning.

"Marigold pigged this afternoon, Mr. Herriot, and I don't think she's got much milk. Little pigs look very hungry to me. Will you come?"

I groaned my way out of bed and downstairs and through the long garden to the yard. By the time I had got the car out into the lane I had begun to wake up and when I rolled up to the inn was able to greet Mr. Worley fairly cheerfully.

But the poor man did not respond. In the light from the oil lamp his face was haggard with worry.

"I hope you can do something quick. I'm real upset about her—she's just laid there doing nothin' and it's such a lovely litter. Fourteen she's had."

I could understand his concern as I looked into the pen. Marigold was stretched motionless on her side while the tiny piglets swarmed around her udder; they were rushing from teat to teat, squealing and falling over each other in their desperate quest for nourishment. And the little bodies had the narrow, empty look which meant they had nothing in their stomachs. I hated to see a litter die off from sheer starvation but it could happen so easily. There came a time when they stopped trying to suck and began to lie about the pen. After that it was hopeless.

Crouching behind the sow with my thermometer in her rectum, I looked along the swelling flank, the hair a rich copper red in the light from the lamp. "Did she eat anything tonight?"

"Aye, cleaned up just as usual."

The thermometer reading was normal. I began to run my hands along the udder, pulling in turn at the teats. The ravenous piglets caught at my fingers with their sharp teeth as I pushed them to one side but my efforts failed to produce a drop of milk. The udder seemed full, even engorged, but I was unable to get even a bead down to the end of the teat.

"There's nowt there, is there?" Mr. Worley whispered anxiously.

I straightened up and turned to him, "This is simply agalactia. There's no mastitis and Marigold isn't really ill, but there's something interfering with the let-down mechanism of the milk. She's got plenty of milk and there's an injection which ought to bring it down."

I tried to keep the triumphant look off my face as I spoke, because this was one of my favourite party tricks. There is a flavour of magic in the injection of pituitrin in these cases: it works within a minute and though no skill is required the effect is spectacular.

Marigold didn't complain as I plunged in the needle and administered 3 c.c. deep into the muscle of her thigh. She was too busy conversing with her owner— they were almost nose to nose, exchanging soft pig noises.

After I had put away my syringe and listened for a few moments to the cooing sounds from the front end I thought it might be time. Mr. Worley looked up in surprise as I reached down again to the udder. "What are you doing now?"

"Having a feel to see if the milk's come down yet."

"Why damn, it can't be! You've only just given t'stuff and she's bone dry!"

Oh, this was going to be good. A roll of drums would be appropriate at this moment. With finger and thumb I took hold of one of the teats at the turgid back end of the udder. I suppose it is a streak of exhibitionism in me which always makes me send the jet of milk spraying against the opposite wall in these circumstances. This time I thought it would be more impressive if I directed my shot past the innkeeper's left ear, but I got my trajectory wrong and sprinkled his spectacles instead.

He took them off and wiped them slowly as if he couldn't believe what he had seen. Then he bent over and tried for himself.

"It's a miracle!" he cried as the milk spouted eagerly over his hand. "I've never seen owt like it!"

It didn't take the little pigs long to catch on. Within a few seconds they had stopped their fighting and squealing and settled down in a long, silent row. Their utterly rapt expressions all told the same story—they were going to make up for lost time.

From: *All Creatures Great and Small*, by James Herriot.

Unusual Patients

Buster Lloyd-Jones

All kinds of unusual animals arrive at a vet's surgery and some of them come to stay. Buster Lloyd-Jones was not short of unusual patients.

One afternoon a tortoise arrived with a broken jaw. He had tangled with a motor mower in Sir Colin Anderson's magnificent garden in Hampstead and I had to pin his jaw with wire and feed him with crushed lettuce and vitamins from a fountain-pen filler. He was a pleasant patient fellow, who became extraordinarily friendly and tame. He stayed on too after he was well and made his home in my garden.

He would plonk himself on my chest when I was sunbathing and would go charging off after other tortoises throughout the mating season. We would hear him making violent love to them—clank, clank, clank. Tortoise shells don't make for discreet relationships. After a particularly noisy night we found him dead on the lawn. His heart had given out. Even a tortoise can have too much of a good thing.

There were other patients who would have been happy to stay too, but they wouldn't have fitted so easily into the family.

Take George, the "largest porcupine in captivity." This is what Brighton Zoo said he was and they may

have been right. He certainly was a big porcupine. One day he escaped from his cage and, in doing so, damaged his nose. Two keepers chased after him and naturally he whirled round and shot a quiverful of quills with deadly accuracy into their legs. Off they went to hospital and the zoo called me in.

Diana Abbott and I found George entangled in the wire netting of a disused tennis court. He was frightened, edgy, in pain and ready to give battle.

Helped by a long stick and a lot of luck we managed to coax him into a box and whisk him off to Dene's Close. Then, of course, our problems had just begun.

The porcupine's nose was lacerated and he wasn't letting anyone near it. So he became the only patient I ever had to approach from behind a dustbin lid. I held it up like a gladiator's shield in one hand, a long stick with cotton-wool soaked in liquid garlic, nature's disinfectant, in the other, edging nearer and nearer. George watched warily. Every time he rattled his quills in warning I would jump out of the way. One quiverful of quills narrowly missed me, rattling on the dustbin lid and making me feel like a medieval knight.

Eventually, with lots of time and patience, I got near enough and somehow managed to clean up the wound. The worst was over. After that he didn't mind so much and as the days went by he got to like us. He would watch out for our coming and give little grunts of welcome. And by the time he got back to the zoo he was literally eating out of our hands.

Then there were the wolves. There were four of them in the zoo and they all caught a kind of wolvine distemper.

I treated them, installed them in a kennel in my isolation ward, did my rounds and went to bed. I was awakened by frightening screams from one of the kennel maids. It had been her day off and, returning late, she had looked into the kennels to see how her patients were. She walked in, expecting friendly dogs, and was petrified to find herself facing a wolf pack. She fled and so did the wolves.

The rest of the night was chaotic, but for the wolves a great spree.

They broke into the chicken farm next door and rampaged among them, enjoying an orgy of slaughter and terror. Dachshunds got in to join in the fun, and at the end of it all I was left to foot the bill for the slaughtered chickens.

I've always advocated natural feeding for wolves as much as anyone else. But my neighbours' chickens weren't exactly what I had in mind.

One of the wolves, more ambitious than the rest, came to a violent end that night. Leaving his companions to the chickens, he set off in the direction of London, where, as is well known, the pickings are easier for a wolf of daring and resource.

Alas, he met a taxi head-on in the London Road and was killed outright. The taxi driver, thinking he had hit an Alsatian of some kind, must have been astounded to find he had run over a Canadian timber wolf in the heart of Sussex.

But of all his adopted patients it was perhaps Pollyo the parrot and Wanda the monkey who most engaged the affections of Buster Lloyd-Jones.

It was at this time that two of my closest friends came into my life.

Pollyo came first.

My assistant at that time, Valerie Higgins, had been in charge of Pets' Corner in the London Zoo and still had many friends there. One day she told me that one of the Amazon parrots had bitten his keeper and was under sentence of death.

I immediately got in touch with the zoo. Could I have the parrot? Certainly, they said. As soon as you like.

So two days later Valerie arrived with Pollyo in a cage.

He was a magnificent old bird, about sixty-five years old, a vivid green with bright, alert eyes, and we all admired him immensely.

"According to the keeper," said Valerie, "he was given to the zoo by the Prince of Wales."

"He's a vicious one," said the keeper. "He'll get his teeth into you if he can."

From Pollyo came a cackle of derisive laughter and then he began to swear. He swore most foully and with great expression.

"The Prince of Wales?" said my mother faintly. "Indeed!"

One thing the keeper had been very sure about was Pollyo's character. He had impressed on Valerie the need for great caution. On no account were we to let Pollyo out of his cage. He was a very dangerous bird.

That night, when everyone was in bed, I softly played the gramophone, carefully opened the cage and sat on a nearby chair.

Pollyo came to the door, looked around, hopped on to the table, waddled across and flopped down on my knee. I sat perfectly still while he made an inspection of my trousers, my jacket, my tie. Gradually he climbed on to my shoulders and gave my face a thorough inspection. He explored my nostrils and my ears and my mouth. He carefully examined each of my eyelashes, looking into each eye, raising and lowering the lids with his beak. Satisfied, he relaxed and from then onwards became part of the family.

He was particularly fond of Wanda and Wanda of him. This was very proper. Both, after all, were reformed juvenile delinquents. Both had been under sentence of death. They had much in common.

Wanda was a Javanese monkey from a private monkey house a few miles away. The owner was desperately short of food for the monkeys and some had to go.

Now I had always wanted a monkey, so I got myself there as quickly as I could.

Wanda was sitting glowering in her cage. One look at me and she went mad, hissing and grimacing and jumping with rage. I loved her on sight.

"She's a vicious one," said her keeper. "She'll get her teeth into you if she can."

Wanda bared her teeth and scowled and gibbered and her amber eyes flashed fear and hatred. I thought her very beautiful.

We got her into the cage we had brought without being bitten, though Wanda did her best, and I drove her blissfully back home. I was the one who was blissful. She seemed to hate every minute of it.

I kept her cage in the surgery so that I could spend as much time as possible with her. Gradually Wanda and I got to know each other and the great day came when I felt I could let her out. I opened the cage door —and at that moment was called to the phone. I was only gone a few seconds but that was long enough for Wanda.

Immediately she swung herself on to the shelves which were laden with my hard-come-by stock of emulsions, linaments and medicines and seized on them with joy, hurling bottle after bottle against the opposite wall.

By the time I got back, my surgery, always fanatically neat and white, was a shambles, the floor littered with broken glass, the walls dripping with my precious medicines.

She saw me and redoubled her efforts, chattering away with the thrill and excitement of it all.

I was bursting with genuine fury. "WANDA!" I bellowed, slamming my hand on the desk.

She froze with surprise. She had never heard me raise my voice before and her face puckered with astonishment. Then, swiftly, she swung down from the shelf and into my arms making funny little "er-er-er" noises in her throat.

From that moment she was my slave and constant companion. We walked hand in hand to visit the dogs kennelled in the stables and she rode on my shoulder into the village. She loved to watch operations and would peer knowingly into dogs' mouths and ears as she had seen me do. Visitors to my surgery had no difficulty in telling which of us was the vet. I was the one in the white coat.

From: *The Animals Came in One by One*, by Buster Lloyd-Jones.

Winged Pets

Gerald Summers

Owned by an Eagle

Buster Lloyd-Jones's Pollyo was unique, but there's nothing unusual, of course, in having a parrot as a pet. Some extraordinary birds, though, have been adopted by humans—and in turn have adopted them. The most extraordinary of all, perhaps, was the ten-week-old eaglet acquired by Gerald Summers in 1960 and given the name of Random. Under his care the eaglet grew into a fine specimen of golden eagle—often described as the most magnificent of all wild birds. Summers trained Random and then let her fly free in the parks of east London and above the Hampshire coast. Always she came back to him. Then he took her to the mountains of Wales.

Somewhat light-headed with the sense of elation that had overtaken me and with little thought for the consequences, I slipped Random's leash from her jesses and held her aloft. With an upward lift of her wings and a powerful back kick of her great yellow feet, she was away. To watch a golden eagle sweeping above the gentle domesticated fields of the English south country is impressive enough, but to see the same bird in action against the backdrop of the Welsh mountains is another thing altogether. Random seemed to double in stature, to take on an air of controlled majesty that even I, who had known her in all her moods, had never fully realised she possessed.

Perhaps my perception was heightened because it occurred to me at that moment that this possibly was to be the parting of our ways. If it was to be so, I wouldn't have had it otherwise. The choice was hers and hers alone. We had been together for nine years, not all that long perhaps, as one judges time, but it seemed an eternity to me at least. She had come to be an integral part of my life. What she thought of me is pure conjecture but judging from the way she behaved, her greeting of me after my absences, and the fact that she had always returned to me of her own free will, I was confident that, despite her limited power of expression, the love and respect I felt for her was returned in full. This belief in her affection and feeling of comradeship was now to be tested.

As she slanted out, swinging far beneath the shadow of the hillside before climbing into the upper air, the sun caught and gilded the metallic feathers on her nape and hackles, turning them to richest copper. Her primaries caught and fingered the rising air currents, toying delicately with the gentle breeze. Half closing her wings she hurtled earthwards; then, checking her fall, once more she swung out over the valley which dropped away almost vertically below her. For the first time I had the experience of looking down on her as she cruised in wide circles, glorying in the freedom of her natural environment.

Her telescopic eyes must have noted the distant forest below, with its hint of wild game to be chased and seized at will. The pair of buzzards towering on high saw this immense bird, a bird the like of which they had never seen before; only the occasional heron, laboriously flapping in a straight line from fishing ground to heronry, had anything approaching this interloper's width of wing-span, and the buzzards knew the herons well, knew that they were inoffensive anglers offering no threat. This was a different proposition: a bird like them in form and outline but with nearly double the wing-span and five times the weight, a bird that by its very presence offered a nameless menace.

The big broad-winged hawks' plaintive mewing call took on a sudden hint of savagery, as, with the larger female leading, they closed their wings and fell out of the sky with a hissing stoop of which even a peregrine might not have been ashamed. Whether the indignant hawks intended to press home the attack is impossible to say. Despite its humble diet of rodents and carrion the buzzard is no coward. Even the raven, its equal in size and its superior in armament, although ever ready to mob the deceptively leisurely-looking raptors, is wary of getting within reach of the buzzard's short, stubby, but exceedingly powerful talons, preferring, by taking advantage of its masterly wing-power, to chivy its slower rival until in sheer exasperation it

takes refuge in a tree. Yet despite this apparent unwillingness to engage in close combat, I have noticed that, if a buzzard is intent on going somewhere specific, it goes there, irrespective of the gang of noisy black hooligans at its tail.

Random was enjoying herself, giving herself up to the sheer pleasure of the moment as she breasted the broad bosom of the strengthening wind which bore her up, playing with her as if she were a piece of drifting gossamer. Listening to the air humming through her slotted primaries she did not at first hear the rushing descent of the approaching foe. Something, instinct perhaps, made her glance upwards. The female buzzard was only about fifty feet above her and was coming down like a six-inch shell, closely followed by the male. Random, never slow to accept a challenge, had no intention of shirking this one. Apparently standing on her tail, she went up vertically to meet the enemy head on. The buzzards must have realized in that instant that here was an opponent such as they had never encountered before, but with their weight and the speed of their descent they were unable to pull out of their headlong dive. Just when a collision seemed inevitable and Random, nimble as a matador, had thrown herself on her back to take the brunt of the attack, some lightning reflex action came into play. As a rushing mountain torrent divides and by-passes a boulder in mid-stream, so the two buzzards by-passed Random; even as her clutching feet shot out to grapple they were safely past and continuing earthwards with undiminished speed. As the grassy hillside rushed up to meet them they levelled out and swept into the branches of a solitary oak.

The whole performance had taken only seconds but it was a magnificent display of natural artistry faultlessly carried out by all three participants. Random righted herself and, once more on an even keel, looked round to see what had become of the opposition. Finding herself alone in the sky she banked sharply, waggling her wings in what looked to me very much like a modified version of the victory roll, and slanted downwards in one long gentle gliding movement. She landed within a few feet of where I was standing enthralled. Random had proved once again that, despite all the obvious attractions of this intriguing new country, with all its hazards and challenges, she still, at least for the present, preferred to throw in her lot with me.

From: *Owned by an Eagle*, by Gerald Summers.

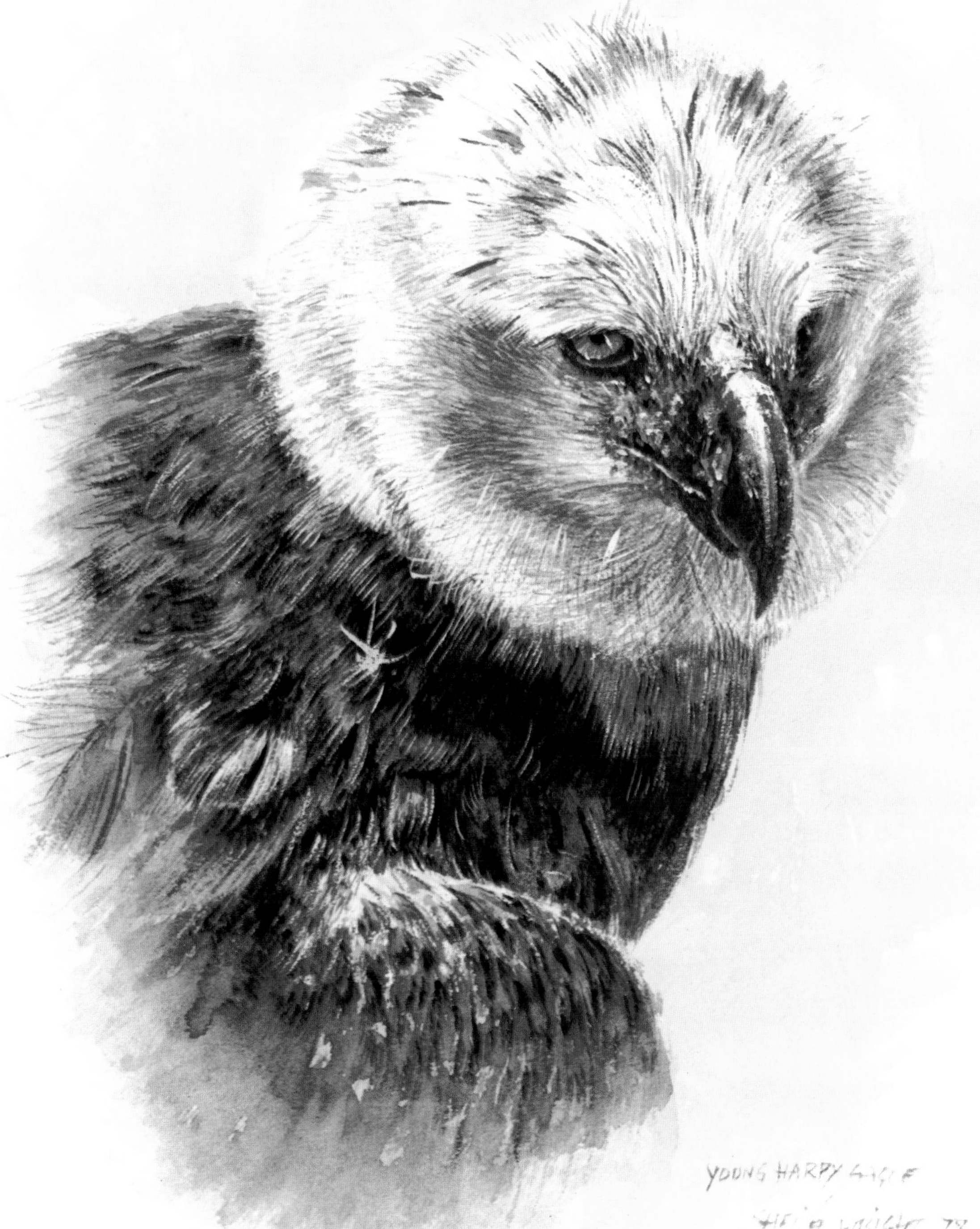

HARPY EAGLE

A

SHARK

B

The Lure of the Falcon

Gerald Summers' introduction to wild birds came as a boy when he found and kept a young falcon. The most extraordinary part of his life with the falcon, Cressida, was during the war when he managed to keep Cressida with him on his various postings in Britain—and she was still with him when his unit was sent to North Africa. They both survived the battles that followed, but eventually Gerald and his comrades were captured by the Germans, and he found himself in a single-decker bus, converted into an ambulance, on the way back to the enemy lines:

Opposite me was a German soldier, half of whose face was swathed in blood-soaked bandages: we looked at each other but did not speak. There were more shouts outside. A uniformed driver climbed into the driver's seat and after a good deal of revving and snorting the machine bounded forward and we were off. Try as I might I could not keep awake. I knew I was slipping into a sort of coma, which I fought hard to overcome, but sheer weariness and a sort of despair combined to produce a powerful anaesthetic. Before I passed out I felt for Cressida where she rested in my blouse. I stroked her head, and she gave my finger a reassuring tweak, which was the last thing I felt before I blacked out.

When I came to I found that I had slipped forward and was lying with all my weight on the German soldier opposite me. I pulled myself together and returned to a more or less upright position. The soldier, still without speaking, produced a bar of the most delicious chocolate I have ever tasted. Suddenly I realised I was famished, having eaten nothing since the previous evening. I finished the bar and made appropriate noises to express my thanks. My opposite number then produced a packet of cigarettes and offered me one. I noticed they were a well-known American brand, but forbore to comment on this. The ambulance, if that is what it was, lurched and thundered through the night. I slept fitfully until we reached the hospital which was just outside Bizerta. Here the ambulance stopped and we all either climbed out or were carried out.

I was taken under escort to an interview room which again seemed full of Germans, including an interpreter and the MO of the hospital, a major in the Medical Corps. I was asked various questions, then the order came which I had been dreading ever since my capture: "Take off your tunic." I explained that

owing to wounds it was impossible, and a number of uniformed figures stepped forward and slit off what was left of my sleeve.

I noticed a window was open in a corner of the room and I made a desperate plan. I put my left hand slowly into my blouse and pressed it gently against Cressida's breast. I felt her grip it with her claws and felt her weight as she climbed on to it. As the medical orderlies began to ease the now sleeveless jacket off I felt for the end of the shortened jesses and took a firm grip with my fingers. After a struggle, like that of a large moth emerging from a very tight chrysalis, the jacket was eased over my head. There in the lamp-light, with her great lambent eyes calmly scrutinising the foe, sat Cressida.

I raised my head and waited for Nemesis to strike. Had there been, as I expected, a bellow of wrath and a concerted rush to take her from me I intended to try to reach the window and cast her off to take her chance. I knew that her wing would soon be strong enough to bear her up; she was in perfect condition, as fat as butter, and if she had had to fast for a few days it would have done her little harm. Besides, with the abundance of grasshoppers, crickets and large beetles that I knew were there for the catching, even a half-witted kestrel should have been able to make some sort of a living, and Cressida was far from being half-witted. But the expected tirade never came.

I glanced at the doctor who was staring at Cressida with an expression of rapture. At last he spoke. "Ah," he said, "eine Turmfalke." Thus I learned the German name for a kestrel. The atmosphere in that small crowded room relaxed at once. The doctor, through the interpreter, asked me various questions about Cressida, whom he thought must be an official mascot. I told him her whole story. The interpreter's English was not brilliant and my German almost non-existent, but we somehow got the whole thing across. The doctor, it seemed, was a keen amateur naturalist, and he had been a practising falconer before the war; he even produced from his wallet a somewhat faded photograph of a splendid goshawk to prove it. Thus it came about that the only British POW with a tame kestrel was confronted with probably the only practising falconer in the German Medical Corps in North Africa. Truly the ways of Destiny are strange and wonderful.

The doctor took Cressida on his ungloved left fist, the correct one, and she for her part behaved impeccably: than he put her on the back of a chair and they busied themselves with me. I was a mess, but after the assorted pieces of metal had been extracted from my shoulder and chest, hot water, disinfectant and surgical stitches soon put me to rights. I took Cressida once more on my fist and, escorted by the now friendly and almost respectful bodyguard, I was taken to a ward and shown my bed—a real bed with white sheets and pillow-cases, a bed such as I had not seen since I left Britain nearly three months before. A copy of *Deutscher Arbeiter Zeitung*, or something similar, was produced and spread on the floor between the wall and the back of my bed, and Cressida was installed upon the brass rail at the head after a piece of bandage had been wrapped round it to protect her feet from the chill.

I climbed thankfully into bed, a medical orderly produced some kind of painkiller or sedative, and after swallowing it I became engulfed in the deepest sleep I have ever experienced. The last thing I remember was the shadow of Cressida, hugely magnified by the flickering lamps, and the great central table in the ward. She seemed to take on the stature of an eagle as she watched over me, surprisingly self-confident and at ease, as I was dragged down into the depths of a coma born of sheer mental and physical exhaustion.

When I surfaced hours later I just could not realise what had happened. I suppose I was still semi-drugged, for the whole situation seemed unreal. However, the truth was soon brought home to me by Cressida, who was by now extremely hungry and waiting for me to do something about it.

The German OC, the major I had encountered the previous night, arrived to make his round of inspection. He walked down the row of beds prescribing treatment for the various injuries. When he came to me he discussed my case with his retinue, spoke a few words to me and to Cressida and, just as he was about to move on to the next bed, produced a small tin box and handed it to me. It contained three plump, dead mice. I was lost for words. Cressida's feeding problem was seen to for the day at least. She ate two mice in rapid succession, as much at home as if she were sitting on a pollard willow in the middle of Romney Marsh. Soon after her meal she descended from her perch and gave herself a sort of imaginary dust bath on my bed, shuffling about and wagging her tail like an old hen. This also showed that she was in a thoroughly relaxed and contented frame of mind, and ready for anything that might come her way.

From: *The Lure of the Falcon,* by Gerald Summers.

Havelock Ellis—the Story of a Brief Friendship

Spike Milligan

Spike Milligan has specially written this true story just for us. It is one of the most affecting stories in our collection—the more so as it contrasts with Spike's outward image as a zany comic.

On the night of March 25, 1943, D. Subsection of the 19th Battery 56 Heavy Regiment, Royal Artillery, moved its giant 7.2 howitzer from Bou Arada to the village of Munchar. It was a cool North African spring night, the sky with stars-a-plenty. When we arrived, the advance party had found a half-bombed house as a billet for the two officers and three signallers. We set up our No. 19 Field Wireless Set and tuned into the Regimental Network.

By dawn the gun had been dug in, lines of coordination observed, and ready to shoot. As the sun grew it revealed our surroundings. It was, or rather had been, a French colonial farming village of some eighteen homes, set in the shadow of a great white upthrust rock called Djbel Munchar. Shelling and dive bombing had reduced the village to rubble. The building that housed us bore signs of hurried evacuation, the saddest item, a little rag doll with its legs burnt off. The roof had been removed by shellfire, but the first floor acted as a roof to the ground floor

where we made ourselves comfortable, as only soldiers can. Our job was to be decoy gun. Every night, we moved to a new position, fired a few rounds, raced somewhere else and repeated the performance. The idea was to confuse the Germans as to how many guns, and where they were.

It was on the evening of the third day. In the gathering dusk, about one hundred yards away, I saw what I thought at first was a wolf. Standing in the tall dry wheat grass, I could only see the top half of his body. Wolves? I didn't remember reading there were wolves in North Africa. I called to Lieutenant Budden, "Excuse me sir, have a look at this." By the time he arrived the "wolf" had slunk off. "There are no wolves in North Africa, Milligan" said Lieutenant Budden, looking at me very strangely. Perhaps I was seeing things. Next evening about the same time, there it was again, this time Lieutenant Budden saw him as well. "It must be a dog" he said. It was.

The following week he appeared again and again, always at twilight. Through binoculars he looked a cross between Alsatian and border collie. I whistled to him several times and he pricked up his ears, but never came nearer. That night we heard that unmistakable sound of a dog sniffing under the front door. I opened a window and flashed my torch. There was the dog. He stared menacingly into the beam, and backed away with a threatening growl. I doused the torch and he was gone. Who was he? I decided to win the animal's friendship. There is an ancient desire in man that challenges him to make friends with the wild beast, using trust as his only barter. Every evening from then on I waited for him to appear. I would then make a display of holding up a chunk of beef, after which I threw it towards him, the gesture would frighten him away, but next morning, the food was always gone.

I continued this for a week. On the last day, he didn't shy away. After a preliminary sniffing of the air, he loped forwards to where the bully was, pausing every five paces or so. I kept stock still. Reaching the meat, he took it, and ran into the approaching dark. I felt a glow of success. Next stage was to get him closer. This I did by dropping the food nearer and nearer to myself. It took ten days but it worked! I got him coming within ten feet of me, but there he stopped. Try as I may, he would not come closer. So, a second plan. I'd starve him for three days, then put the food right beside me. So as not to break the rhythm of our meetings, I appeared every evening but laid no food. He would come forward to look for it. It nearly broke my heart to see his thin body, looking desperately, frantically, for the food, but I didn't give in. The fourth evening, this was the test. I threw the food five feet away from me. He watched me, then, instead of his hesitant halting approach, he moved forwards rapidly, he stopped just once to look at me, then quickly ran forwards, snatched the food, ran some thirty yards, turned, looked back at me, and was gone. I'd done it! But God! one needed patience.

There were other drawbacks: our own guns were deafening, and the return German shellfire spoiled many meetings with the dog. Apart from being hungry, the animal was terrified, so you will understand my feelings of success whenever I gained a step nearer to winning his confidence. There was one major holdup—I was sent up to the observation post on Djbel Munchar mountain. I made provision for Gunner Alf Fildes to leave the food out every night, till I returned. *If* I returned.

The week at the O.P. dragged by. I found myself worrying about the dog, like a mother over a sick child. Over the wireless I received a daily message assuring me all was well. The evening I returned, the sky over Africa opened up, and rain fell like a Niagara from heaven. The roar as it hit the earth drowned out all the sounds of guns. Visibility was nil, but, donning my rain cape, I went out to continue the man and animal tryst. It occurred to me that here I was, in the middle of a world war, and I was finding this evening

meeting with a half-wild starved dog of paramount importance. I slopped through the red mud towards where I usually met him. By sheer accident we stumbled on each other. His fur clung to his body, showing how really emaciated he was. In the brief moment before he ran away, I noticed he wore a collar. He must have been a French farmer's dog, Arab dogs never wore one.

The next five days saw him edging closer, until finally he came three feet from me, but he took nearly a whole ten minutes to do it, every second of which his eyes were fixed on me. For the first time, I spoke to him, using any schoolboy French I could remember and keeping my voice in a coaxing tone. He, in turn, kept the whisper of a growl going deep in his throat, as though to say "This is not my real frightening growl, but just watch out." Now I would try and feed him by hand. I didn't know what I had let myself in for. Half an hour of holding your hand at arm's length, in the squatting position, and keeping up the flow of insane cooing noises had me giving up, but, it finally paid off. Late one evening he threw caution to the wind and snatched the food off my hand. That momentary feel of his warm muzzle in the palm of my hand was an indescribable moment. In these barren war surroundings, it brought a moment of sanity— it brought back those sunlit early summer days, with the many dogs that had given me joy in my boyhood, Trixie, Boxer, Laddie. All came back in that one moment.

The next move was to try and stroke him, but, it wasn't me who succeeded! It was Lieutenant Budden, who had taken a fatherly interest in the whole affair. He was 31 and very English. In pre-war days he was a professor of English and mathematics. He wore iron frame glasses, looked permanently puzzled, and without fear. One evening as the dog came to feed, Budden walked straight to him, patted his head and said, "Hello darling, is darling hungry?" The dog was obviously as stunned as I was, but Budden's method worked. "We've got to give him a name Milligan, and I just happen to have one."

Looking down at the dog he solemnly intoned, "I name this dog Havelock Ellis." So it was. Havelock took to sleeping at the side door of the house. I made up a rough kennel, placed an old blanket inside, a spare old tin for his dinner bowl, and a mug for his water. During the days, he wandered off, but at sundown, he always returned to his kennel and stayed there at night. What a watch dog! *No* one, except the five of us who lived in the house, could get near the place. To date we noticed that he had never wagged his tail. He was very much like a child who'd had a tough childhood. He seemed grateful, but not happy. As dogs go he was quite big, knee-high with a thick brown and black coat. Once again, Lieutenant Budden worked a miracle. Returning from a drive, to Headquarters, I heard joyous barking from the house. Inside, I saw Lieutenant Budden holding a biscuit above Havelock's head and the dog was jumping at it, but, most important of all, he was wagging his tail! We didn't know then that Havelock had only two days to live.

He would come in the house quite freely now—early morning he would lick the face of the sleeping forms of my fellow soldiers. He was now a happy dog again, but he still had mental scars. He got plenty of love. We risked our lives and gave him a bath. He loved it when we groomed him with a boot brush. Then came that night.

It was very dark, and very quiet. There were a few distant sounds of guns, firing to the left of our line. We had blacked out the windows, and lit our variety of lights from candles to oil-filled cigarette tins. Budden was reading aloud excerpts from John Donne as we were playing cards. Suddenly outside we heard Havelock let loose a most savage growl, and race to the front of the house. First thoughts were

"Germans!" We doused the lights, felt for our tommy guns. Next, the shout of a man being attacked mixed with the savage snarling from Havelock. A shot, an agonized yelping, then silence. By now we had run out the back of the house and made our way to the front. I went to where Havelock lay. There was a bullet hole in his throat. He died as I patted and held his head, but before he did, he managed one last wag of his tail.

The intruder was a man wearing blue denims and a beret. He had a pistol. We disarmed him, and took him inside. In broken English he told us the story.

This was his house. He, his wife and three children had been evacuated after the village was shelled. This evening he had returned to find his dog, which had run away when the house had been hit. It was Havelock, his own dog, that he had shot in the darkness when it attacked him. In self-defence he had shot him.

Next morning, with no word spoken we dug a grave, using an empty twenty-five-pound charge case. We lined it with his blanket, and buried him. On a post I wrote a simple epitaph.

The fact that thirty-six years after his death I've written this story shows I will never forget him.

Hats Off to Cats

Buster Lloyd-Jones

Buster Lloyd-Jones has had some fascinating experiences with Siamese cats:

Pandora was a Siamese cat who thought she was a dachshund.

She came to me extremely sick, sent by people I didn't know and couldn't trace. Later I found that a judge had given her to his son and daughter-in-law. They had gone off on holiday, leaving the poor thing locked in an expensive flat in Belgravia with a few bottles of milk with the tops off and some opened tins of cat food. They were away for weeks and when the cat was found at last she was in a pitiful state.

When she came to see me she was so ill that for four weeks she would neither eat nor drink. The stamina of animals can be astonishing. She became a skeleton, poor creature, unable to move, but she was still alive.

We gave her crushed garlic, honey and water, as much sun as possible and a lot of love, and very, very slowly she began to take an interest in life again. All her life she had been cooped up in a small flat and had seen few, if any, dogs before. Now she attached herself to a paralysed dachshund and shared his basket and it was touching to see the love these two sick animals had for each other.

The owners never got in touch with me again, thinking, no doubt, she was dead, and so she stayed. She grew into a sweet, lovely creature but to the end

of her life she clearly believed she was a dachshund herself and only seemed really happy with dachshunds round her. We never had the heart to tell her.

Soon afterwards two more Siamese cats joined us. Their names were John and Josephine. They came to me after the divorce of their owners. They had gone with the wife who adored them—but she had had to go back to work and couldn't look after them. She asked if I would keep them for the time being. I agreed, and they duly arrived. Before she left, their owner gave elaborate instructions about food and exercise to the kennel maid and left with words that became famous at Dene's: "They'll be all right," she said, "just as long as they never meet a dog."

Well, of course you couldn't be at Dene's Close for five minutes without meeting dogs, much to the pleasure of John and Josephine. They were, in fact, more like dogs than cats. They played with the dogs for hour after hour and at night they slept in their kennels.

I had both for the rest of their lives—John for nine years and Josephine for twelve. They were always together but they were, as actresses say to airport reporters, just good friends.

Where love was concerned Josephine ignored poor John completely. She preferred a neighbouring ginger tom whom we all called Ginger Rogers. She produced litter after litter—and always the father was Ginger. I was always hoping for at least one pure-bred

Siamese litter, if only as a boost for John's ego, but it was Ginger who had the sex-appeal.

Every now and then Josephine got asthma—sometimes quite badly. She would wheeze away and go very thin and look quite dreadful. I would be able to ease the discomfort, but asthma is a tricky thing to deal with and Josephine had these attacks all her life.

Just before Christmas one year a school teacher who lived nearby called in for some advice. She had found this pathetic Siamese cat, she said, absolutely starving. She seemed to be in a very bad way. What should she do? I gave her advice about feeding the cat up and putting it back into condition with powders and vitamins and she went her way.

A few days later she was back to borrow a cat basket. She was going to Exeter for Christmas, she said, and she would have to take the poor cat with her.

That night I couldn't find Josephine anywhere. She had been in and out of the surgery, wheezing with asthma all morning, but no one had seen her since.

Suddenly the penny dropped. The teacher's starved Siamese was asthmatic Josephine, who had got on to a good thing. She had eaten her usual hearty dinner at Dene's and had popped out for another hearty meal across the garden.

I rang the teacher at once. Too late. She had already left for Exeter, taking Josephine with her. I was a bit worried about Josephine's asthma and anyway we wanted her home for Christmas, so I got the teacher's holiday address and rang her up.

"You know that half-starved Siamese?" I said. "Well, her name is Josephine and she belongs to me. Could I have her back, please?"

Josephine arrived back next day. We put the basket on the surgery table, opened it and out she jumped looking marvellous. Her asthma had passed and she was sleek and handsome and delighted to be home again. She purred and rubbed herself against us, was unusually affectionate to John and then made for the kennels to sleep it off.

From: *The Animals Came In One by One*, by Buster Lloyd-Jones.

Oakhill Albert Apricot and Others

John Newell

John Newell is a breeder—and of course a lover—of cats. He has recounted some of his experiences for us.

It is all very well to say that a tom cat should be allowed to live a free natural life, but usually it isn't practical. The outcome is either endless kittens forced upon every fertile female cat in the vicinity or else keeping him locked up, yowling and spraying and making the night and day and atmosphere hideous. The alternative is the sex life of the professional stud cat, snobbish and commercialised to the human, romantic and varied to the stud. It was my own Cornish Rex stud cat, Oakhill Albert Apricot, who first introduced me to the unexpected complexities of feline behaviour.

Bert, as he was known to his intimates, was the first-born of a litter of three—there was no doubt about the order since his mother, Briarry Sweety Tart, usually produced her kittens in the middle of the bed and liked to have people in attendance. Like most first-born kittens Bert was very extrovert and not very intelligent (it is the last-born of a litter who is, usually, introverted, original, intelligent but frequently very neurotic). As he reached puberty Bert developed a fixation on his sister—Oakhill Victoria Plum—and spent most of the day trying to mate her. Any other female was savagely attacked. Fortunately, Bert was psychoanalysed by a friend, a cat breeder of great experience, who gave him a mild anaesthetic and put him in a cage with a randy and experienced female. As the mists of anaesthetic cleared, Bert found himself the subject of confidently amorous attentions and was soon mating to the manner born.

Then he settled down to the routine of the professional stud. A new female arrived for him each week in a wicker basket and, after a cautious introduction, the two were left to get on with it. But etiquette—and Albert's stud fee—demanded that some matings at least were witnessed. That's when I learnt that the phrase "making love like animals" is meaningless—at least as far as cats are concerned, though I suspect the bestial couplings of dogs really are what we call animal. Every affair was different. One old girl, whose owner had said she would mate "with anything"—not very complimentary to Albert—retired to the top of the bookcase and stayed there for a week, nor during that time did Bert show the slightest interest in her. A young, small, virgin female whose owner left her with tears in her eyes, extracting repeated promises of Bert's extraordinary gentleness and chivalry, completely exhausted him. The same experienced breeder advised us to separate them before he had a heart attack—an occasional cause of death in the young stud. There was something terrifying in the sight of the little female frisking around Bert's prostrate body, anxiously licking and talking and asking for more. They slept in one another's arms and fed from the same dish.

But before Bert could become established as a stud he had to gain some recognition as a pedigreed animal carrying the desirable, hereditable characteristics of his breed. And this meant he had to be shown, competing in cat shows to win enough prizes to enable a suitable advertisement ("Oakhill Albert Apricot, Stud") to be placed in *Fur and Feather*, the official organ of the Cat Fancy—the feline equivalent of the

Kennel Club. Bert didn't like being shown. It meant getting up at six in the morning—an hour when Bert was usually still just a roundish splodge on the middle of the bed—and being polished all over with a chamois leather, then being put in a wicker cat basket which he detested and driven through the early morning streets to some gaunt Victorian hall or other. There he would stand in a long, cold queue for hours with upsetting rival tom cat smells drifting down the queue and finally have his basket opened and be poked about by a vet. Bert would emerge on such occasions like a demon king, with his ears completely flattened to his skull, looking (as my wife once said) like a "Chinese earless cat" and making a strange and terrifying crooning noise wavering up and down the scale like the wind in the chimney of a lonely house. The vets usually nodded him through to his cage where he lost valuable points by sulking all day. But being a handsome ginger animal and well typed, as they say, Bert did well enough and was soon travelling to shows all over the country.

I remember once taking Bert to a show in Bournemouth (held in another draughty monument to the Gothic Iron Age). We went by train, in a long open carriage. Bert had gotten used to brief trips in his basket in cars, but trains were something new. The train was crowded and as I squeezed down with the basket perched on my knee the noise began. It was a thudding, bestial yowl interspersed with sounds of splintering wickerwork. After a very few moments a crack opened under the straining lid and a paw with splayed claws thrust through and jerked threateningly about. The noise got worse. A small child sitting next to me crawled silently away on to its mother's knee. I tried to push Bert's paw unostentatiously back into the basket but the only effect was to redouble the yowls. Bert had a streak of Siamese, always evident at times of stress. The muttering in the carriage grew louder. I could pick out a few words now, such as "Shameful," "Disgusting" and ". . . ought to tell the guard." Bert's little pink nose, somehow bearing a completely unwarranted look of terrified pathos, emerged for a moment from the relentlessly widening gap between lid and basket. I rammed it back with the flat of my hand, but not before the lady opposite, who had been working up her indignation for some time, was finally driven to speak. "Can't you let the poor little chap out?" she said.

The trouble was that, once Bert was out in an open carriage, there would not be the slightest chance of getting him back in again. A parallel situation is said to exist with genies and bottles. The lady opposite didn't know this but I did. By this time Bert had his head and half shoulder out, which meant that, short of sitting on the lid and squeezing him into a marmalade pulp, there was no way of stopping the rest from emerging within a few moments. I snatched up the basket and ran for the lavatory.

Once inside with the door locked I sat down on the seat and opened the basket. The noise stopped as though it had been switched off. Bert ambled out stretching himself and purring and settled on my knees. I wished I'd brought something to read; "Rock Rules OK" and other, less printable, slogans soon lost their charm.

The contemplative peace in the lavatory was eventually broken by a rising crescendo of whispering and shuffling in the corridor outside. I'd ignored several people trying to get in and muttering about the time some people, etc. But this was different. After a while the whispering died away and instead there was a firm, authoritative rapping on the door. Bert, who had gone to sleep, his preferred activity in the presence of friends and in the absence of food or female cats, ignored it. So, for as long as possible, did I but, when the tapping rose to a hammering pitch with cries of "what's happening in there" I had to reply. Unfortunately there is something about lavatory cubicles which instils a residual sense of guilt in every Englishman. Instead of asking the crowd to disperse in clear, authoritative terms, I made the mistake of mumbling "what's all the fuss about" in a furtive monotone. "WHAT HAVE YOU DONE WITH THAT CAT" came clearly through the door. "I'M EXERCISING IT" I yelled back, losing my cool. The noise outside the door died away again, Bert slumbered on and there was peace.

But not for long. This time the silence was broken by a male voice of authority. "Excuse me sir" it said. "Am I right in thinking that you have a cat in there with you?" "That is in fact so, yes," I replied, picking up the style. "I shall have to ask you to vacate the toilet immediately" said the voice—clearly that of the guard. "This is contrary to British Rail's regulations."

The basket was open, silence reigned, Bert snoozed. I snatched up the soft, furry bundle, jammed it in, and in the same movement shot the lid down, sat on it, and reached for the straps. But Bert wasn't having any of that! He had rather liked the lavatory—he had an odd habit, common to many cats, of actually drinking water from flush toilets if not prevented—and it only took him a moment to wake up. The cry which rose to trouble deaf heaven outdid any of his former efforts as I emerged shamefacedly, cat basket in hand, and made my way through a hostile crowd back towards my seat. One lady even suggested that the train should be stopped and a, vet fetched "to examine the poor animal." As the train drew into Bournemouth I leapt out and bolted for a cab.

The guard was probably right about B.R.'s regulations. The carrying of cats in English trains is surrounded by a mass of strange, Victorian regulations. I remember once on a country station a lady and myself just catching a train with a cat in a basket, and breathlessly arriving at the barrier. The ticket collector shook his head. "It'll have to be weighed" he said.

I wasn't particularly surprised or daunted myself. I'd learnt to expect anything when transporting cats and this cat was at least a quiet and elderly Siamese. But the lady with me was furious. "Don't be such a fool" she said. "We'll never catch the train." The official shook his head stolidly. "That cat is not being carried in arms," he said. "It is in a basket and must be weighed." We went into the ticket office as the train drew out of the station. The next one was due in two hours. The cat was solemnly removed from the basket and placed on the scales. The official recorded its weight in laborious crabbed copperplate on an appropriate form. A light drizzle began to fall on the exposed platform. My companion snatched the cat from the scales, shoved it into the basket and headed for the platform. "Name of owner," intoned the official. "Ghengis Khan" came from the wet and windy platform. The official's back stiffened but his pen scratched imperturbably on. At last he'd finished and the completed form was presented to me. No one asked for it at the barrier when we finally got to Waterloo Station. It wasn't until weeks later, cleaning out my pockets that I found the piece of paper and the owner's name space filled in as "Mr. G. Khan."

Bert's mother Sweety and later her daughter Victoria Plum introduced me to a warm and passionate world, that of the mother cat and her litter. Kittenhood is all too short; in nine or ten weeks the mother is losing interest—though just occasionally love lasts for life, as with Sweety and Plum who would place all their combined litters in one box and then form a sort of living trough around them for suckling whilst purring symphonically. Cat littermates can indeed remain devoted friends for life, as can mothers and daughters or sons, but the rule is otherwise. As Muriel Beadle, an authority on cats, says "Year-old cats behave to their mother or their littermates as they do to unrelated cats. Mother mates with son, daughter with brother, brothers fight each other. There is no memory of clustering together to pool their warmth, no remembrance of the mice their mother brought them. Now they too walk by themselves." But why does a cat on its owner's knee, at its most ecstatic, "make bread" with its feet, purring and kneading with eyes shut in deep delight? Because it is reliving its untroubled infancy, when the kneading movements of the front paws helped to stimulate the flow of mother's milk. "The human speaks to the remains of the childish impulses and so revives them," says Dr. Paul Leyhausen, another student of cat behaviour. "Thus it is possible to have a lasting friendship between man and cat."

I've come to believe that the fact that human and adult cat are the same relative sizes as cat and infant kitten is part of the basis for the bond. Close to the warm human, a cat can return thankfully to the tragically brief few weeks of kittenhood, of warmth and milk and all-enveloping furry protection. If such vulnerable regression and abandonment is the basis of the deepest bonds between cats and humans, then it is perhaps not surprising that, at other times, our cats balance it by walking through life aloof and independent, by themselves. And what a happy privilege for the cat owner, to be able magically to bring back the golden age to the cold and lonely hunter, the adult cat.

That Mad, Bad Badger

Molly Burkett

Badgers can also make delightful pets—adorable in many ways, but a great trial in others. Even their virtues can be vices, as Molly Burkett describes in her hilarious account.

Badgers are noted for their tidy habits and Nikki was no exception to this rule. In the wild, badgers will keep their nests in a spotless condition. Anything out of place will be removed and discarded. Even a leaf that has been blown into the entrance hole, or a piece of straw that has been dropped when they renewed their bedding, will quickly be taken out.

Nikki developed this tidiness habit to an extreme. A piece of string, a bit of mud, anything that offended her sense of neatness was quickly gathered and hidden in one of her hiding places. At first Mum thought it was a delightful habit and one that the rest of the family should follow. She soon changed her mind about that though, because as Nikki increased in size so did the number of things that she considered out of place. Cushions, chair-backs, even cutlery from the table would be gathered up and hidden in one of her favorite places in an untidy heap. I can remember Mum replacing the chair-backs six or seven times one morning before she gave it up.

The picture was much the same upstairs. Mum would have to crawl under the bed to recover her make-up, and probably her nylons and her under-clothes as well if she had forgotten to close the chest of drawers up tightly.

One day when Aunt Gerty had come to tea, Nikki made a real entrance. She must have found a drawer a little open, Mum was always forgetting to shut them, and she had made the most of the opportunity. She suddenly exploded on us, completely entangled in Mother's underwear. Somehow she had got her head through a bra of Mum's and the rest of the garment was trailing along behind her and kept tripping her up, and each time she tripped she became more and more entangled. Aunt Gerty's face was a study.

You can have no idea how tidy our house looked when Nikki lived in it. Dressing tables were cleared regularly, as was every other surface which held ornaments or the like. A trail of mauled flowers up the staircase would let us know that Mum's newest flower arrangement was the latest thing to upset Nikki's sense of order. A wail from Sophie would inform us that the badger had had similar feelings about the way she had laid the table and had neatly cleared it all off again.

Cushions, library books, potted plants, anything that could be moved proved irresistible to the badger. If I put my homework down for a couple of seconds or Dad put his cigarette packet down, they would be whisked away before we had time to realise they had gone.

I shall never forget the day that Nikki attempted to discover whether cigarettes were more than articles to tidy away. We found her sitting on the stairs, half way up, an open packet of cigarettes beside her. She had stuck one into the side of her mouth and was using those capable forepaws to turn the cigarette round as if she was trying to find out what Dad really found to appreciate in them.

Handbags were another badger delight. Most of our women visitors would deposit their handbags casually on the floor beside them. That really did offend Nikki's tidiness cult and the bag would be hurried away almost as soon as it was deposited. There was generally a great panic when the disappearance was discovered and Mum would have to make the rounds of Nikki's hiding places in order to restore the bag to the owner.

That was until the affair of Aunt Gerty's bag.

Aunt Gerty believed in outsize handbags. Dad reckoned she used to carry everything in hers except the kitchen sink. She certainly did produce the most peculiar things from it from time to time. One day Mum could not find the tin opener and Aunt Gerty dug into her voluminous handbag and produced one from its depths. Another time she found a cheese sandwich in it when she was looking for something quite different. Dad reckoned the sandwich was so old it was mummified.

Aunt Gerty always plumped herself down in the easychair Nikki preferred. On this occasion she started talking as usual while her anatomy was still poised in mid air, as it were, and, as usual, she deposited her bag beside her.

Even I was surprised at the speed of the following sequence of events. Sophie was the first to realise something was amiss. She started to giggle and we soon saw why. Nikki had come in. I cannot say that she was intelligent enough to know how to put lipstick on properly but she certainly had a good try. There were red streaks all over her mask. She knew she had done wrong and tried to worm her way up on to my lap. In the same instant Aunt Gerty discovered that her bag was missing.

Need I say more? Not only had Nikki taken the bag, her nimble paws had discovered how to undo the catch and she had really gone to town on the contents. Three weeks later we were still recovering Aunt Gerty's belongings from all over the house.

Mum began to think things were getting a bit too much when she discovered she could not start preparing Sunday lunch until she had crawled under the bed to retrieve the saucepans from the jumble of miscellaneous objects beneath it and found the sausages she had bought for supper there as well. But what finally incensed her was the way Nikki helped her to do the gardening.

Mum had spent the whole day preparing the flower beds, putting in the plants and some special dahlia corms that she had bought. You can well imagine what she thought (and said) when she went up to her room to change and, looking under the bed for her make-up, found everything that she had just planted neatly laid out in three rows.

Evidently they had offended Nikki's sense of tidiness in some way. She must have been digging them up and carrying them off as quickly as Mum had been planting them.

From: *That Mad, Mad Badger*, by Molly Burkett.

ANIMALS WILD

It's worthwhile, reflecting a little on the term "the wild." In its purest sense it should presumably mean those regions where wild animals roam virtually undisturbed by man, or at any rate where man is merely one small part of the total ecological pattern, taking his place among the other predators but posing no serious threat to the natural balance of the species.

These regions, which once covered a considerable proportion of the globe, are now rapidly shrinking as the human populations soar and vast new areas are brought under cultivation to feed them, and as the needs (actual or imagined) of modern industrialized man make ever-increasing inroads into the natural environment.

Inevitably this has brought about changes in attitude towards the animals which live in the rapidly dwindling wild. These changes have not all been for the bad. There are still plenty of people who exploit wild animals whenever they get the chance, or who regard them as entirely expendable in the pursuit of their own aims. But there are some at least who are seriously and practically concerned about them. In the past the wild was often a playground for the sportsman and its animals were slaughtered indiscriminately. This doesn't mean that sportsmen were all wicked or cruel, of course. There was so much of the wild and so many animals in it that it just didn't occur to most of them that there was any need to set a limit to their sport. At that time the idea of any sense of kinship existing between civilized man and savage animals would (except in a few rare individuals) have seemed utterly preposterous.

Villagers, too, helped deplete large numbers of native species. For example, the beautiful serval, a wild cat common to the brush and grasslands of Africa south of the Sahara, was hunted for its skin, which was used to make mantels for African villages.

Even those travellers who weren't primarily sportsmen thought nothing of shooting the wild animals they encountered, either for food or simply because they *were* wild. They tended to regard the wild as a place apart, an unfortunate adjunct to civilization, an impediment in the way of progress. It was a place of constant danger and of thrilling adventure, and its animal denizens were alien pests, to be treated as natural enemies and eliminated wherever possible. That is why in reading the accounts of nineteenth-century travellers, many of them outstandingly good and brave men, it often comes as quite a shock to us today to see how calmly they could describe their apparent brutality towards wild animals and proudly list their tally of slaughter.

Sometimes, of course, the animals *were* the natural enemies of man. When lions were numerous in South Africa, for example, they were frequently a serious threat—as David Livingstone, the famous missionary and explorer, found when he was at Mabotsa on one of the sources of the Limpopo (or Crocodile) River in 1842.

Caught by a Lion

David Livingstone

It is well known that if one in a troop of lions is killed the others take the hint and leave that part of the country. So the next time the herds were attacked, I went with the people, in order to encourage them to rid themselves of the annoyance by destroying one of the marauders. We found the lions on a small hill about a quarter of a mile in length, and covered with trees. A circle of men was formed round it, and they gradually closed up, ascending pretty near to each other. Being down below on the plain with a native schoolmaster, named Mebalwe, a most excellent man, I saw one of the lions sitting on a piece of rock within the now closed circle of men. Mebalwe fired at him before I could, and the ball struck the rock on which the animal was sitting. He bit at the spot struck, as a dog does at a stick or stone thrown at him; then leaping away, broke through the opening circle and escaped unhurt. The men were afraid to attack him, perhaps on account of their belief in witchcraft. When the circle was re-formed, we saw two other lions in it; but we were afraid to fire lest we should strike the men, and they allowed the beasts to burst through also.

If the Bakatla had acted according to the custom of the country, they would have speared the lions in their attempt to get out. Seeing we could not get them to kill one of the lions, we bent our footsteps towards the village; in going round the end of the hill, however, I saw one of the beasts sitting on a piece of rock as before, but this time he had a little bush in front. Being about thirty yards off, I took a good aim at his body through the bush, and fired both barrels into it. The men then called out, "He is shot, he is shot!" Others cried, "He has been shot by another man too; let us go to him!" I did not see any one else shoot at him, but I saw the lion's tail erected in anger behind the bush, and, turning to the people, said, "Stop a little till I load again." When in the act of ramming down the bullets I heard a shout. Starting, and looking half round, I saw the lion just in the act of springing upon me. I was upon a little height: he caught my shoulder as he sprang, and we both came to the ground below together. Growling horribly close to my ear, he shook me as a terrier dog does a rat. The shock produced a stupor similar to that which seems to be felt by a mouse after the first shake of a cat. It caused a sort of dreaminess, in which there was no sense of pain nor feeling of terror, though quite conscious of all that was happening. It was like what patients partially under the influence of chloroform describe, who see all the operation, but feel not the knife. This singular condition was not the result of any mental process. The shake annihilated fear, and allowed no sense of horror in looking round at the beast. This peculiar state is probably produced in all animals killed by the carnivora; and if so, is a merciful provision by our benevolent Creator for lessening the pain of death.

Turning round to relieve myself of the weight, as he had one paw on the back of my head, I saw his eyes directed to Mebalwe, who was trying to shoot him at a distance of ten or fifteen yards. His gun, a flint one, missed fire in both barrels; the lion immediately left me, and, attacking Mebalwe, bit his thigh. Another man, whose life I had saved before, after he had been tossed by a buffalo, attempted to spear the lion while he was biting Mebalwe. He left Mebalwe and caught this man by the shoulder, but at that moment the bullets he had received took effect, and he fell down dead. The whole was the work of a few moments, and must have been his paroxysm of dying rage. In order to take out the charm from him, the Bakatla on the following day made a huge bonfire over the carcase, which was declared to be that of the largest lion they had ever seen.

It is interesting to note that the effects of paralysis and stupor following the lion's bite and his violent shaking of the victim as described in David Livingstone's vivid account, have been confirmed by later scientific observers.

Land Lizards of the Galapagos Islands

Charles Darwin

There were, of course, plenty of professional, scientific observers of wildlife during the nineteenth century—as you will see from other chapters in this book. The most important of them all, because it was he who profoundly affected man's whole attitude towards the animal creation, and man's own place in it, was Charles Darwin.

In 1831 (when he was only twenty-two years old) he was appointed as naturalist to *HMS Beagle*, which was setting out on a long survey (lasting nearly five years) of the Atlantic and Pacific Oceans. It was the careful and detailed observations of all kinds of animals in their natural habitats—and, above all, of the variations between closely related species on islands and the nearest mainlands—which Darwin made during this expedition that started him thinking about the modification of species brought about by different environments and conditions, and eventually led him to the exposition of his theory of evolution in his world-shaking book *On the Origin of Species by Means of Natural Selection, or the Preservation of Favored Races in the Struggle for Life* (to give it its full title) which was published in 1859.

During the voyage of the *Beagle*, Darwin was particularly interested in the Galapagos Islands, an archipelago in the Pacific, lying along the equator, some 650 miles to the west of the South American country of Ecuador. The name of the islands is derived from the Spanish *galápago*, a tortoise, and refers to the giant forms which had evolved there. Some of these tortoises have reached a length of four feet and more, a weight of nearly 400 pounds, and in individuals attained ages of 300 to 400 years—making them the oldest living animals on earth—but they are practically extinct now, largely due to the carrying away of thousands of them by whalers for food; to the slaughter of great numbers for their oil; and to the wholesale devouring of the eggs by wild dogs and pigs (originally introduced by men).

The most remarkable feature of the fauna of the Galapagos Islands, in fact, and one that excited Darwin, is the large proportion of forms peculiar to the islands—no less than 96 per cent of the reptiles, for example, it is claimed.

Here is a characteristic example of the kind of observations Darwin made while he was in these islands:

Like their brothers the sea kind, they are ugly animals, of a yellowish orange beneath, and of a brownish red color above: from their low facial angle they have a singularly stupid appearance. They are, perhaps, of a rather less size than the marine species; but several of them weighed between ten and fifteen pounds. In their movements they are lazy and half torpid. When not frightened, they slowly crawl along with their tails and bellies dragging on the ground. They often stop, and doze for a minute or two, with closed eyes and hind legs spread out on the parched soil.

They inhabit burrows, which they sometimes make between fragments of lava, but more generally on level patches of the soft sandstone-like turf. The holes do not appear to be very deep, and they enter the ground at a small angle; so that when walking over these lizard-warrens, the soil is constantly giving way, much to the annoyance of the tired walker. This animal, when making its burrow, works alternately the opposite sides of its body. One front leg for a short time scratches up the soil, and throws it towards the hind foot, which is well placed so as to heave it beyond the mouth of the hole. That side of the body being tired, the other takes up the task, and so on alternately. I watched one for a long time, till half its body was buried; I then walked up and pulled it by the tail; at this it was greatly astonished, and soon shuffled up to see what was the matter; and then stared at me in the face, as much as to say, "What made you pull my tail?"

They feed by day, and do not wander far from their burrows; if frightened, they rush to them with a most awkward gait. Except when running down hill, they cannot move very fast, apparently from the lateral position of their legs. They are not at all timorous: when attentively watching any one, they curl their tails, and raising themselves on their front legs, nod their heads vertically, with a quick movement, and try to look very fierce: but in reality they are not at all so; if one just stamps on the ground, down go their tails, and off they shuffle as quickly as they can. I have frequently observed small fly-eating lizards, who, when watching anything, nod their heads in precisely the same manner; but I do not at all know for what purpose. If this *Amblyrhynchus* is held and plagued with a stick, it will bite it very severely; but I caught many by the tail, and they never tried to bite me.

The Komodo Dragon

David Attenborough

With the shrinking of the wild and the disappearance of so many of the species that once belonged to it, it is not easy nowadays to imagine the feelings of excitement and wonders expressed by an observer like Charles Darwin, in the presence of so much that was new and undiscovered. The nearest approximation to these feelings in our own times, perhaps, is to be found in David Attenborough's various *Zoo Quests*. The most exciting account was that of the Komodo Dragon.

This is a much bigger species of lizard than the one described by Charles Darwin—it is in fact the largest lizard in the world. Komodo is a small island (twenty-two miles long and twelve miles wide) which lies almost in the middle of the Indonesian archipelago. Komodo is the lizard's main habitat, though specimens have also been found on two nearby islands. But it doesn't exist anywhere else, and this is something of a mystery. It is almost certainly the descendant of the even larger prehistoric lizards, whose fossil remains have been found in Australia, and the most ancient of these are probably about sixty million years old. But the island of Komodo is a volcanic one of comparatively recent origin, and that obviously raises the questions as to why the "dragons" should exist there, and there only, and how they got there. No one seems to know the answers for certain.

For many years, in fact, the existence of the giant lizards was a matter of hearsay and legend. It wasn't until 1910 that a Dutch army officer (until 1950 Indonesia was part of the Dutch empire, known as the Dutch East Indies) confirmed the stories by shooting two of the creatures and bringing back their skins to Java, where he presented them to the Dutch zoologist Ouwens. It was Ouwens who published the first description of the lizard, and gave it the scientific name *Varanus komodoensis*.

Later expeditions reported that the lizard was carnivorous, living on the flesh of the wild pig and deer which also lived on the island; and that although it usually fed on rotting carrion, it probably hunted its prey as well, killing it with a swing of its huge tail.

David Attenborough's expedition had as its primary objective the filming of the Komodo dragon. Here is his description of the climax of the quest—as he, his companion Charles Lagus, and their Indonesian hunter Sabran, after sending the rest of their helpers back to the boat, waited hopefully near the bait which they had laid.

∾∾∾

There was now little noise. A jungle cock crowed in the distance. Several times a fruit dove, purple-red above and green below, shot with closed wings like a bullet along the clear channel above the stream-bed, soundless except for the sudden whistle of its passage through the air. We waited, hardly daring to move, the camera fully wound, spare magazines of film beside us and a battery of lenses ready in the open camera-case.

After a quarter of an hour, my position on the ground became extremely uncomfortable. Noiselessly, I shifted my weight on to my hands, and uncrossed my legs. Next to me, Charles crouched by his camera, the long black lens of which projected between the palm leaves of the screen. Sabran squatted on the other side of him. Even from where we sat, we could smell only too strongly the stench of the bait fifteen yards in front of us. This, however, was encouraging for we were relying on this smell to attract the dragons.

We had been sitting in absolute silence for over half an hour when there was a rustling noise immediately behind us. I was irritated; the men must have returned already. Very slowly, so as not to make any noise, I twisted round to tell the boys not to be impatient and to return to the boat. Charles and Sabran remained with their eyes riveted on the bait. I was three-quarters of the way round before I discovered that the noise had not been made by men.

There, facing me, less than four yards away, crouched the dragon.

He was enormous. From the tip of his narrow head to the end of his long keeled tail he measured a full twelve feet. He was so close to us that I could distinguish every beady scale in his hoary black skin, which, seemingly too large for him, hung in long horizontal folds on his flanks and was puckered and wrinkled round his powerful neck. He was standing high on his four bowed legs, his heavy body lifted clear of the ground, his head erect and menacing. The

line of his savage mouth curved upwards in a fixed sardonic grin and from between his half-closed jaws an enormous yellow-pink forked tongue slid in and out. There was nothing between us and him but a few very small seedling trees sprouting from the leaf-covered ground. I nudged Charles, who turned, saw the dragon and nudged Sabran. The three of us sat staring at the monster. He stared back.

It flashed across my mind that at least he was in no position to use his main offensive weapon, his tail. Further, if he came towards us both Sabran and I were close to trees and I was sure that I would be able to shin up mine very fast if I had to. Charles, sitting in the middle, was not so well placed.

Except for his long tongue, which he unceasingly flicked in and out, the dragon stood immobile, as though cast in gun-metal.

For almost a minute none of us moved or spoke. Then Charles laughed softly.

"You know," he whispered, keeping his eyes fixed warily on the monster, "he has probably been standing there for the last ten minutes watching us just as intently and quietly as we have been watching the bait."

The dragon emitted a heavy sigh and slowly relaxed his legs, splaying them so that his great body sank on to the ground.

"It seems very obliging," I whispered back to Charles. "Why not take his portrait here and now?"

"Can't. The telephoto lens is on the camera and at this distance it would fill the picture with his right nostril."

"Well, let's risk disturbing him and change lens."

Very, very slowly Charles reached in the camera case beside him, took out the stubby wide-angle lens and screwed it into place. He swung the camera round, focussed carefully on to the dragon's head and pressed the starting button. The soft whirring of the camera seemed to make an almost deafening noise. The dragon was not in the least concerned but watched us imperiously with his unblinking black eye. It was as though he realized that he was the most powerful beast on Komodo, and that, as king of his island, he feared no other creature. A yellow butterfly fluttered over our heads and settled on his nose. He ignored it. Charles pressed the camera button again and filmed the butterfly as it flapped into the air, circled and settled again on the dragon's nose.

"This," I muttered a little louder, "seems a bit silly. Doesn't the brute understand what we've built the hide for?"

Sabran laughed quietly.

"Iss very O.K., tuan."

The smell of the bait drifted over to us and it

occurred to me that we were sitting in a direct line between the dragon and the bait which had attracted him there.

Just then I heard a noise from the river-bed. I looked behind me and saw a young dragon waddling along the sand towards the bait. It was only about three feet in length and had much brighter markings than the monster close to us. Its tail was banded with dark rings and its forelegs and shoulders were spotted with flecks of dull orange. It walked briskly with a peculiar reptilian gait, twisting its spine sideways and wriggling its hips, savoring the smell of the bait with its long yellow tongue.

Charles tugged at my sleeve, and without speaking pointed up the stream-bed to our left. Another enormous lizard was advancing towards the bait. It looked even bigger than the one behind us. We were surrounded by these wonderful creatures.

The dragon behind us recalled our attention by emitting another deep sigh. He flexed his splayed legs and heaved his body off the ground. He took a few steps forward, turned and slowly stalked round us. We followed him with our eyes. He approached the bank and slithered down it. Charles followed him round with the camera until he was able to swing it back into its original position.

The tension snapped and we all dissolved into smothered delighted laughter.

From: *Zoo Quest for a Dragon*, by David Attenborough.

The Pride of Lions

For the most part the wild today exists by courtesy of man—in the sense that enlightened governments have set aside areas where animals can roam in their natural state, protected as far as possible by game wardens and other officials from interference. The vast size of some of these areas, however, really do justify the term "the wild"—and their inclusion in this chapter.

Think, for example, of that vast stretch of semi-arid thornbush in the Northern Province of Kenya, comprising some 120,000 square miles and stretching from Mount Kenya to the borders of Ethiopia. It is an area that is wild indeed, even if it is also patrolled by game wardens. The senior game warden is George Adamson, and his wife, Joy, is the authoress of *Born Free*, its sequel *Living Free*, and of many other animal classics.

Joy Adamson is one of the outstanding instances of an attitude towards the wild which is utterly at variance with that which commonly prevailed in the nineteenth century. For her and others like her, it is a place that *belongs* to the wild creatures that inhabit it, and humans are privileged observers. It is separate, existing in its own right, but no longer to be regarded with fear and hostility; and the approach towards it is, generally speaking, realistic and objective, recognizing the differentness of animals without any sentimental attempt to endow them with cozy human qualities, but doing so with an unpatronizing respect and esteem.

Not that this new attitude excludes the possibility of a close bond between humans and wild animals, as in the case of that between Joy Adamson and the lioness Elsa. It is significant, though, that there was no question of turning Elsa into something *less* than a wild animal by putting her into a zoo or making her into a domestic pet. On the contrary, the Adamsons, attached though they were to Elsa, went to tremendous pains to reintroduce her to the wild as soon as possible, because they knew that it was there that she had her true being.

Their main anxiety now was whether Elsa's contact with humans had made it impossible for her to live in her natural state. Perhaps she had been turned into a creature doomed to inhabit a limbo between the human and the animal world, and perhaps her own kind, sensing the human presence, would turn on her and kill her. To the great relief of the Adamsons, however, Elsa succeeded in readapting herself to the wild, mated and gave birth to three cubs (whom the Adamsons later christened Jespah, Gopa and Little Elsa).

What happened next was in many ways the most amazing part of the whole story, and those of you who have read *Living Free* will remember that unforgettable moment when news was brought to Joy Adamson, while she was working in her studio on a riverbank, that Elsa was calling from the other side of a river, "in a very strange voice." Needless to say, Joy Adamson immediately left her studio and ran along the riverbank until she came to a spot close to the camp, where (in the dry season) a sandbank rose out of the water, not far out from Joy Adamson's side of the river. And there, standing on the sandbank and looking straight at her, was Elsa, with one of the cubs beside her, another just scrambling out of the water, while the third was still on the far side, pacing up and down and "calling piteously."

Now Joy Adamson had been frequently warned that after the birth of her cubs Elsa would almost certainly be fierce and dangerous. So she stood absolutely still. She watched as the lioness, with a soft, moaning sound, went up to the cub which had just emerged from the water and was still shaking itself, and began licking it. Then she lifted her head, looked across the river at the cub still stranded on the far bank, and began swimming back to it, accompanied by the other two cubs.

When the family was reunited, they all settled down in the shade of a fig tree which grew out of some rocks and boulders near the water's edge. Joy Adamson went on watching them from her side of the river. At first the three cubs were suspicious of her presence and hid themselves. But after a while they began peering at her through the undergrowth, and then, overcome by curiosity, came out into the open and stared across the water at her, while Elsa made little humming noises, as if reassuring them that there was nothing to fear from the silent watcher. Then, tired of staring, the cubs began playing with their mother, climbing over her, rolling about on her body, and making grabs at her gently twitching tail.

A little later Elsa got up and went to the river's edge, as if contemplating another crossing. Meanwhile Joy Adamson had given orders for a carcass of meat to be brought. When the boy arrived with it, Elsa flattened her ears and looked fierce, but as soon as Joy Adamson was alone again, Elsa and one of her cubs swam right across the river, and Elsa began tearing at the carcass. But her cub returned to the water—perhaps because it was still frightened, or perhaps because he meant to help the other two cubs across. When Elsa saw that it had entered deep water she plunged in after it, caught up with it and, grabbing its head in her mouth, gave it a good ducking, as if punishing it for being too venturesome. She then carried it in her mouth back to Joy Adamson's side of the river, and a little later another cub plucked up

its courage and swam across, though the third, more timid one, still stayed where it was, meowing plaintively.

This time Elsa went close to Joy Adamson and began rolling on her back. It seemed to Joy that the lioness was trying to teach her two cubs that this human was to be trusted—was in effect one of the pride. Very cautiously the two cubs began to creep closer, until they were about three feet away. At this point Joy had to restrain a natural impulse to lean forward and touch the cubs, because she knew that the initiative must be left entirely to them. Any premature move on her part would have panicked them and undone all Elsa's efforts at effecting an introduction. In fact, on this occasion the cubs obviously felt that three feet was the proper limit of their approach.

Joy Adamson left them alone for about an hour, though still staying at her post. Then she called out to Elsa, who at once replied, in the kind of voice she had always used in talking to her human friend—a quite different sound from that which she used with her cubs. Attended by the three cubs Elsa now walked down to the water's edge—and this time all of them swam across.

When they had landed, Elsa carefully licked each of her cubs in turn. Then she turned towards Joy. In the past her habit had been to charge up to her mistress when she came out of the water. But this time Elsa walked up to Joy slowly, rubbed herself gently against her, rolled over on her back, then got up again, licked her face and hugged her—and again Joy had the feeling that the lioness was trying to teach her cubs that they, too, could become Joy's friends. But although the cubs watched with great interest, they still looked puzzled and stayed out of reach.

Elsa now took the cubs over to the carcass, and they began licking and tearing at the skin in great excitement. Joy Adamson calculated that they were a little over six weeks old, so this may have been their first experience of a "kill." She was now able to study them more closely, noting that although they still had a blueish film over their eyes (not unlike human babies) they could see perfectly, and that they were in excellent condition, with beautiful glistening coats. She wasn't able, at this stage, to ascertain their sex, but she noticed that the cub with the lightest coat was the most attached to its mother, cuddling up close to her whenever it had the opportunity, preferably under her chin, and catching at her with its paws.

By now it had grown dark, and Elsa took her cubs some distance away into the bush, and a moment later Joy heard the sounds of suckling. So she gradually made her way back to the camp—to find Elsa and the cubs waiting for her no more than ten yards from her tent!

Now came what was perhaps the biggest surprise of all. When Joy went to her tent, Elsa followed her. Once inside the tent, she threw herself on the ground and called out to her cubs to come and join the party. Although they did venture up to the tent, they wouldn't go in, but stayed outside, meowing. So Elsa went out to them. Joy Adamson followed her. She sat down on the grass. And Elsa came and leaned against her, while her cubs nuzzled up close to their mother, searching for her teats. Soon, Joy Adamson tells us, as the moon rose and the nearby palm trees were silhouetted against its light, there was no sound to be heard in the African night except for the suckling of the cubs.

A human being, a lioness, reputedly ferocious and untrustworthy in motherhood, and three cubs, all contentedly bunched together—it must have been a remarkable tableau. No wonder Joy Adamson tells us that she felt "very humble!"

Based on: *Born Free* and *Living Free*, by Joy Adamson.

A Cheetah's Honor

Somehow it's the big cats one associates with the wild, more than any other animal. Not that the cheetah is by any means the biggest of them. It seldom measures more than four feet in length. By contrast, a lion shot in Uganda in 1969 was just one inch under eleven feet and a Bengal tiger, shot in 1907, measured ten feet, seven inches—and greater lengths have been claimed, especially for the Siberian Tiger, which is now practically extinct. The leopards (or panthers), to whose family cheetahs belong, can also reach a length of well over seven feet, if the tail is included.

Neither is the cheetah the most ferocious member of the cat family, nor the most dangerous to man. That doubtful distinction probably goes to the big leopard (*Pantherus parda*), which is a notorious man-eater. In 1910, for example, one specimen in India accounted for no less than 400 victims, and as recently as 1972 a leopard attacked and killed three boys within the space of eight hours in villages near Junagadh, western India. The individual man-eating record, though, belongs to the "Champawat maneater"—a particularly ferocious tigress which terrorized parts of Nepal and adjacent territories, and killed 438 people in eight years before being shot in 1911. As for lions, they generally prefer to hunt in prides, but sometimes an individual, expelled from its pride for one reason or another (including disease or old age), will become a lone killer of men: in 1963, for example, one of these "rogue" lions killed fourteen people in Malawi within a month.

There are good reasons, however, for thinking of the cheetah as a particularly characteristic denizen of the wild. It is for one thing, a very successful hunter —the alternative name for it is "hunting leopard." Its success is mainly due to the fact that it is the fastest of all the hunters. It isn't the fastest animal over long distances. That, undoubtedly, is the pronghorn antelope of the western U.S.A. In 1918 a group of these antelopes raced parallel with a car on a road in New Mexico at an average speed of 30 m.p.h. over a distance of seven miles, and higher speeds than that have been recorded, especially over shorter distances. But when it comes to short bursts, of the kind needed for surprise attacks on fleeing prey, the cheetah is the fastest animal on earth. Over level ground it can reach a speed of 60 m.p.h., for a distance up to 450 yards, and much higher speeds have been claimed. Some of these are almost certainly exaggerated, but there's no doubt at all that the cheetah is a magnificent sprinter! In addition, its body is extraordinarily sinuous and agile, so that it can twist and turn at top speed. Professor Milton Hildebrand, a zoologist at the University of California, Los Angeles, has worked out that even if the cheetah didn't have any legs, it could still propel itself along the ground at a speed of 5 m.p.h. (and that's a pretty fast human walking pace) because of its streamlined body and its astonishing muscular contractions!

The cheetah is another of the animals of the wild in which Joy Adamson has interested herself. With one cheetah, named Pippa, she established a relationship almost as close as that with Elsa—and in the course of it she learned many fascinating details of wild animal behavior.

One of the most fascinating of all the incidents she has related is that when Pippa, who had just had her fourth litter of cubs, found herself challenged by Mbili, a fully grown daughter from an earlier litter. Now Mbili was defending her own territory; she was the bigger and stronger of the two—and there was no reason to believe that either animal would "remember" their relationship after this span of time. So Joy Adamson was deeply troubled as the two cheetahs faced each other—Mbili determined to defend her territory and Pippa to defend her new cubs, which she had left feeding under a nearby tree. Then, just as Pippa was about to spring, Mbili suddenly rolled over on to her back, in the surrender position, with a low, moaning sound. At once Pippa turned and went back to her cubs.

This act of surrender is not uncommon in the wild. In their fights for leadership both lions and wolves practice it. When a wolf comes to the conclusion that it cannot win a fight, it will turn on to its back, submissively offering its throat, as the most vulnerable part of its body, to the moral victor. Similarly, the lion will offer its soft, unprotected belly. In both cases the victor will never take advantage of this surrender, satisfied that his challenger has admitted defeat—a magnanimity of behavior, one might think, that human beings could well adopt more frequently. Unruly cubs use the same methods of submission to their exasperated mothers. But in this instance neither of these two issues were at stake. And Joy Adamson came to the conclusion that, even in their absence, "a similar code of honor existed between mother and daughter cheetah"—and continued to operate over a considerable period.

It is interesting to note that cheetahs have often been trained by man to hunt small deer, gazelle, hare, foxes, and even certain birds. Cheetahs are now mostly found on the open plains and semi-desert savannahs of East Africa, Iran, Turkey and Afghanistan. Until quite recently they also frequented India, but became extinct (in their wild state) in the early 1950's. Some Indian princes still employ cheetahs for the hunt, but they are imported from Kenya.

Based on: *The Searching Spirit,* by Joy Adamson.

A Beaver's Home

James A. Michener

Popular author James A. Michener includes in his best-selling novel, Centennial, *this warm account of a beaver's struggle to survive.*

When the bison straggled over the land bridge into America he encountered a huge misshapen creature that was in many ways the opposite of himself. The bison was large in front, slight in the rear, while the native animal was very large in the rear and slight in front. The bison was a land animal; the other lived mostly in water. The beast weighed some three hundred and fifty pounds as it slouched along, and its appearance was fearsome, for its conspicuous front teeth were formidable and as sharp as chisels. Fortunately, it was not carnivorous; it used its teeth only to cut down trees, for this giant animal was a beaver.

It had developed in North America but would spread in desultory fashion through much of Europe; its residence in the streams of Colorado would prove especially fortuitous, bringing great wealth to those Indians and Frenchmen who mastered the trick of getting its pelt.

The first beavers were too massive to prosper in the competition that developed among the animals of America; they required too much water for their lodges and too many forests for their food, but over the millennia a somewhat smaller collateral strain became dominant, with smaller teeth and softer pelts, and they developed into one of the most lovable and stubborn of animals. They thrived especially in the streams of Colorado.

One spring the mother and father beavers in a lodge on a small creek west of the twin pillars made it clear to their two-year-old daughter that she could no longer stay with them. She must fend for herself, find herself a mate and with him build her own lodge. She was not happy to leave the security in which she had spent her first two years; henceforth she would be without the protection of her hard-working parents and the noisy companionship of the five kits, a year younger than herself, with whom she had played along the banks of the stream and in its deep waters.

Her greatest problem would be to find a young male beaver, for there simply were none in that part of the creek. And so she must leave, or in the end her parents would have to kill her because she was mature enough to work for herself and her space inside the lodge was needed for future batches of babies.

So with apprehension but with instinctive hope, this young female left her family for the last time, turned away from the playful kits and swam down the tunnel leading to the exit. Gingerly, as she had been taught, she surfaced, poked her small brown nose toward shore and sniffed for signs of enemies. Finding none, she gave a strong flip of her webbed hind feet, curling her little paws beneath her chin, and started downstream. There was no use going upstream, for there the building of a dam was easier and all the good locations would be taken.

One flap of her hind feet was sufficient to send her cruising along the surface for a considerable distance, and as she went she kept moving her head from side to side, looking for three things: saplings in case she needed food, likely spots to build a dam and its accompanying lodge, and any male beaver that might be in the vicinity.

Her first quest was disappointing, for although she spotted quite a few cottonwoods, which a beaver could eat if need be, she found no aspen or birch or alders, which were her preferred foods. She already knew how to girdle a small tree, strip its bark and fell it so that she could feed on the upper limbs. She also knew how to build a dam and lay the groundwork for a lodge. In fact, she was a skilled housekeeper, and she would be a good mother, too, when the chance presented itself.

She had gone downstream about a mile when there on the shore, preening himself, she saw a handsome young male. She studied him for a moment without his seeing her, and she judged correctly that he had chosen this spot for his dam. She surveyed the site and knew intuitively that he would have been wiser to build it a little farther upstream, where there were strong banks to which it could be attached. She swam toward him, but she had taken only a few powerful strokes of her hind feet when, from a spot she had not noticed, a young female beaver splashed into the water,

slapped her tail twice and came directly at the intruder, intending to do battle. It had taken her a long time to find a mate and she had no intention of allowing anything to disrupt what promised to be a happy family life.

The male on shore watched disinterestedly as his female approached the stranger, bared her powerful front teeth and prepared to attack. The stranger backed away and returned to the middle of the stream, and the victorious female slapped the water twice with her tail, then swam in triumph back to her unconcerned mate, who continued preening himself and applying oil to his silky coat.

The wandering beaver saw only one other male that day, a very old fellow who showed no interest in her. She ignored him as he passed, and she kept drifting with no set purpose.

As late afternoon came on and she faced her first night away from home, she became nervous and hungry. She climbed ashore and started gnawing desultorily at a cottonwood, but her attention was not focused on the food, and this was good, because as she perched there, her scaly tail stretched out behind her, she heard a movement behind a larger tree and looked up in time to spot a bear moving swiftly toward her.

Running in a broken line, as she had been taught, she evaded the first swipe of the slashing paw, but she knew that if she continued running toward the creek, the bear would intercept her. She therefore surprised him by running parallel to the creek for a short distance, and before he could adjust his lunge to this new direction, she had dived to safety.

She went deep into the water, and since she could stay submerged for eight or nine minutes, this gave her time to swim far from where the bear waited, because even from the bank a bear could launch a powerful swipe which might lift a beaver right onto the bank. When she surfaced, he was far behind her.

Night fell, the time when her family had customarily played together and gone on short excursions, and she was lonely. She missed the kits and their noisy frolic, and as night deepened she missed the joy of diving deep into the water and finding the tunnel that would carry her to the warm security of the lodge.

Where would she sleep? She surveyed both banks and selected a spot which offered some protection, and there she curled up as close to the water as she could. It was a miserable substitute for a proper lodge, and she knew it.

Three more nights she spent in this wretched condition. The season was passing and she was doing nothing about the building of a dam. This bothered her, as if some great purpose for which she had been bred was going unattended.

But the next day two wonderful things happened, the second having far more lasting consequences than the first. Early in the morning she ventured into a part of the creek she had not seen before, and as she moved she became aware of a strong and reassuring scent. If it were serious, and not an accident, it would be repeated at the proper intervals, so she swam slowly and in some agitation to the four compass directions, and as she had anticipated, the keen smell was repeated as it should have been. A male beaver, and young at that, had marked out a territory and she was apparently the first female to invade it.

Moving to the middle of the stream, she slapped her tail, and to her joy a fine-looking young beaver appeared on the bank of the creek and looked down into the water. The slapping could have meant that another male had arrived to contest his territory and he was prepared to fight, but when he saw that his visitor was the kind he had hoped to attract, he gave a little bark of pleasure and dived into the stream to welcome her.

With strong sweeps of his webbed feet he darted through the water and came up to her, nudging her nose with his. He was highly pleased with what he found and swam twice around her as if appraising her. Then he dived, inviting her to follow him, and she dived after him, deep into the bottom of the creek. He was showing her where he intended building his lodge, once he found a female to help.

They returned to the surface and he went ashore to fetch some edible bark, which he placed before her. When beavers mated, it was for life, and he was following an established pattern of courtship. The female was eager to indicate her interest, when she noticed that his gaze had left hers and that any fruitful communication had ended.

He was looking upstream, where one of the most beautiful young beavers he had ever seen was about to enter his territory. This female had a shimmering coat and glowing eyes, and she swam gracefully, one kick sending her to the corners of his areas, where she checked the markers he had left. Contented that she was in the presence of a serious suitor, she moved languidly to the center of the area and signaled with her tail.

The young male left his first visitor and with lightning strokes sped to this newcomer, who indicated that she was interested in the segment of the creek he had laid out for himself and was willing to move in permanently. In this brief space of time their destiny was determined.

What now to do with the first visitor? When the new female saw her she apprehended immediately what had happened, so she and the male came to where the young beaver waited and started to shove her out of the delimited area. But she had got there first and intended to stay, so she dived at the intruding female and started to assault her, but the male knew what he wanted. He had no desire to settle for second best, so he joined the newcomer, and together they forced the unwanted intruder downstream, and as she disappeared, chattering in rage, they slapped their tails at her and made joyous noises and prepared to build their dam.

The outcast drifted aimlessly and wondered whether she would ever find a mate. How could she build a home? How could she have kits of her own? Bitterly she sought the next miserable place to spend a night.

But as she explored the bank she became aware of a soft sound behind her and was certain it must be an otter, the most fearful of her enemies. She dived deep and headed for any cranny within the bank that might afford protection, and as she flattened herself against the mud she saw flashing through the waters not far distant the sleek form of an otter on the prowl.

She hoped that his first sweep would carry him downstream, but his sharp eye had detected something. It could have been a beaver hiding against the bank, so he turned in a graceful dipping circle and started back. She was trapped, and in her anxiety, fought for any avenue of escape. As she probed along the bottom of the bank she came upon an opening which led upward. It could well be some dead end from which there was no escape. But whatever it was, it could be no worse than what she now faced, for the otter was returning and she could not swim fast enough to escape him.

She ducked into the tunnel and with one powerful kick sent herself upward. She moved so swiftly that she catapulted through the surface and saw for the first time the secret cave that had formed in the limestone, with a chimney which admitted air and a security that few animals ever found. Soon her eyes became accustomed to the dim light that filtered in from above and she perceived what a marvelous spot this was, safe from otters and bears and prowling wolves. If she built her dam slightly below the cave and constructed her lodge in the body of the creek, attaching it by tunnel to this secret place, and if she then widened the chimney upward and masked its exit so that no stranger could detect it, she would have a perfect home. To complete her delight she found inside the cave and above the water level a comfortable ledge on which she could sleep that night.

Before dawn she was at work. Moving to all the prominent places on the shore and to the ledges in the creek, she stopped at each and grabbed a handful of mud. With her other hand she reached to the opening of her body where two large sacs protruded and from these she extracted a viscous yellow liquid which would become famous throughout the west as castoreum, one of the most gratifying odors in the natural world.

Kneading the castoreum into the mud and mixing in a few grasses to make the cake adhere, she placed it carefully so that its odor would penetrate in all directions, and when she had set out nine of these—for this was a spot worth preserving and protecting—she stopped and tested the results of her labor. She swam upstream and down, and wherever she went she got the clear signal that this stretch of water belonged to a beaver who intended holding it.

She became, that summer, a capable beaver, lively in her pursuit of things she required. The limestone cavern became not only a place of refuge but also a satisfactory home. She built three long secret escape hatches, one leading a good twenty feet inland from the bank of the creek, so that if a bear or wolf did take her by surprise, she could dive into it and make her way back to her home before the predator knew where she had gone.

The cycle of her life, however, was still incomplete. By herself she would not build a dam, nor a lodge either, for they were needed primarily for the rearing of young. She could survive in the limestone cave, but without the act of building a lodge with a mate, she was still an outcast.

This did not prevent her from attending herself as carefully as ever. Each day, when the sun was low, she perched on the bank overlooking her domain and preened. She did this by using the two peculiar toes on each of her hind legs; the nails on these toes were split so as to form small combs, and these she dragged through her pelt until even the slightest irregularity was removed. Then she took oil from her body and carefully applied it to each part of her coat, combing it in deeply until her fur glistened in shimmering loveliness. No one saw or applauded this grooming, but it was impossible for her to go to bed until she had completed it.

And then, in early autumn when she had given up hope of finding a mate, a shabby beaver seven years old who had lost his family in some catastrophe, wandered down the river and turned by chance into her creek. He was by no means a handsome creature; indeed, he was not even acceptable, for a long gash ran down the left side of his face and he had lost the

two toes on his left hind leg that he needed for cleaning himself, so that his appearance was disreputable.

As he sashayed up the creek he detected the markers and realized immediately that a mistake had been made. The creek spot looked inviting but any flood from the river would wash it away. He looked about for the family which occupied it to warn them of the danger they faced, and after a while he saw the head of the owner breaking through the surface. She swam out to him cautiously and looked for his mate, while he looked for hers. There was a period of motionless silence. He was tired and winter was at hand.

They stared at each other for a long time, for a very long time, and each knew all there was to know. There would be no illusions, no chicanery.

It was he who broke the silence. By the way he looked and moved his tail he indicated that this spot was no place to build a dam.

With a fierce toss of her head she let him know that this was where she would live. And she led him underwater to the entrance of her secret cave and showed him the escape hatches and how she planned to link it to the lodge and the dam, but still he was not satisfied, and when they surfaced, he started to swim to a much safer spot, and she followed, chattering and slamming her tail and halting in disgust as he left her premises.

In the morning he swam back and indicated hesitantly that she was welcome to accompany him if she would consent to build their dam at a proper site.

Again she abused him, protesting furiously and snapping at him, driving him from her water, and that afternoon he came back quietly with a length of aspen in his teeth. Diving to the bottom of the creek, he fastened it to the floor with mud, the first construction in their new home.

It was then September and they set to work with a passion. They labored all night, dragging trees and branches into the stream, weighting them with mud and gradually building the whole construction high enough to check the flow of water. Again and again as they worked he betrayed his doubt that the dam they were building would hold, but she worked with such fervor that he swallowed his precautions.

When the two beavers were satisfied that the dam would impound the water necessary for their establishment, she began tying branches and tree lengths into the bottom, weighting them with rocks and mud and other trees, and it was now that she realized that in the building of the dam she had done most of the work. He was great on starting things, and showed considerable enthusiasm during the first days, but when it came time for doing the hard, backbreaking work, he was usually absent.

She had to acknowledge that she had accepted a lazy mate, one who could not be cured, but instead of infuriating her, this merely spurred her to greater effort. She worked as few beavers, an industrious lot, had ever worked, lugging huge trunks of trees and slapping mud until her paws ached. She did both the planning and the execution, and when the pile from which their lodge would be constructed was nearly finished, and she was eleven pounds lighter then when she started, he indicated for the final time that when the floods came, this would all vanish. She made no response, for she knew that just as she had done most of the building this time, she would have to do it again if floods ever did come.

When the pile in the middle of the small lake behind the dam was completed, they dived to the bottom and began the gratifying task of cutting entrances into it, and providing sleeping levels above the waterline, and places for kits when they came, and digging connecting runways to the secret chamber, and at this planning he was a master, for he had built lodges before.

Only a few days remained before the freeze, and this period they spent in a burst of superenergy, stripping bark and storing it for their winter's food. Where eating was concerned, he was willing to work, and in the end they had a better lodge than any other on the creek, and better provisioned too.

In the early days of winter, when they were frozen in, they mated, and in spring, after she gave birth to four lovely babies, the river produced a flood which washed away the dam and most of the lodge. He grunted as it was happening, but she rescued the babies and took them to higher ground, where a fox ate one.

As soon as the floods receded, she began to rebuild the dam, and when it was finished, she taught the babies how to help rebuild the lodge, which took less effort.

They then enjoyed four good years in their tight little kingdom, but on the fifth, sixth and seventh years there were floods, the last of such magnitude that the whole establishment was erased. This was enough for him, and he spent considerable time upstream looking for a better site, but when he found one, she refused to move. He found her marking the corners of her estate with castoreum and teaching her children how to start erecting a higher and better dam.

He halted at the edge of her territory and watched as this stubborn little creature proceeded with her engineering, making the same mistakes, dooming her dam to the same destruction.

He was now fifteen years old, an advanced age for a beaver, and she treated him with respect, not requiring him to haul logs or do much actual construction on the lodge. He snapped at the kits when they placed branches carelessly, indicating that if he were in charge, he would not accept such sloppy workmanship. As he aged, his face grew uglier, with the scar predominating, and he moved with crotchets and limps, and one day while he was helping girdle some cottonwoods, he failed to detect a wolf approaching and would have been snatched had he not been bumped toward the safety tunnel by his mate.

That year there was no flood.

Then one day in early autumn when the food was safely in and the lodge never more secure, she happened to wander up the tunnel into the secret place which the family had so much enjoyed, and she found him lying there on the limestone ledge, his life gone. She nudged him gently, thinking that he might be asleep, then nuzzled him with affection to waken him for their evening swim through the lake they had built and rebuilt so many times, but he did not respond, and she stayed with him for a long time, not fully comprehending what death signified, unwilling to accept that it meant the end of their long and necessary companionship.

In the end the children took the body away, for it was no longer of any use, and automatically she went about the job of gathering food. Dimly she sensed that now there could be no more babies, no more kits playing in the limestone chamber and scampering down the runways.

She left the security of the lodge and went to each of the compass points and to the salient ridges in between, and at each she scooped up a handful of mud and mixed it with grass and kneaded in a copious supply of castoreum, and when the job was done she swam back to the middle of her lake and smelled the night air.

This was her home, and nothing would drive her from it, neither loneliness nor the attack of otters nor the preying of wolves nor the flooding of the river. For the home of any living thing is important, both for itself and for the larger society of which it is a part.

From: *Centennial*, by James A. Michener.

Colobus Monkey Business

Joy Adamson

There seems to be no end to the animals of the Kenyan wild which Joy Adamson has studied with scientific thoroughness and a deep intuitive understanding. She has specially sent us this account of a very different species from either Elsa or Pippa.

I am sure that you would like to know about the four Colobus monkeys who live in my garden in Kenya. They are completely free but because they know me and trust me they occasionally allow me to feed them with carrots.

The Colobus are the most handsome of all monkeys with their black woolly coats, white long haired capes (mantles) and tail tassels and with their white whiskered and bearded faces crowned by a double capped black bonnet. They differ from other monkeys in that they have only a vestige of a thumb (colobus in Greek means "thumbless"), their stomachs are large and have sections, almost like a ruminant's, to enable them to digest leaves, and they do not have a prehensile tail although infants have one for a short period before the tail tassel develops. The infants are almost white when first born. The monkeys nearest to them in habits are the howler monkeys who live in the Panama Canal Zone. Colobus are strictly limited to the equatorial belt in Africa.

When I first moved into my present home in 1970, I found a pair of Colobus monkeys living in the trees and not long after our arrival they produced a son. I watched them every day as they played in the trees which provided these arboreal monkeys with all the food they needed. I had always thought that a monkey jumped by instinct but I soon learnt that the infant had to be taught to judge the distance it could safely jump by hopping from the shoulder of one parent to the other parent. The parents would sit opposite each other about three feet apart and would gradually widen the distance between them until "Coli," as I called the infant, could safely jump it—if he made a mistake he got spanked! He was also taught how to judge whether branches would bear his weight as he jumped onto them.

The numbers of Colobus are greatly reduced because of poaching and habitat destruction. Not only is the skin of the Colobus very attractive but the Africans consider that their meat is delicious to eat and so poachers prey heavily on them. Not long ago two lorries were stopped on their way to Mombasa—one lorry contained 28,000 Colobus pelts, the other 23,000, all of which would have been illegally shipped to furriers in other countries. Poaching of that order will soon exterminate these good-natured beautiful monkeys.

TIGER

C

(Top) AFRICAN ELEPHANT
(Bottom) INDIAN ELEPHANT AND CALF

D

I had decided that if I could get Colobus monkeys who lived in the wild to adopt young ones who were born in captivity, that we might be able to restock the "bush" with captive born or rehabilitated monkeys and ensure that the Colobus would not become extinct in the wild.

Before I could start this experiment, I arrived home one day to find that Coli's father had been shot by poachers. It was heartbreaking to see the deep distress of Coli, who was now almost full grown, and his mother. For over two months they never played but just cuddled together as they gazed into space, almost as if they were in mourning. Then I obtained two young Colobus females from a film company and from the moment I brought them home Coli and his "mum" took a great interest in them. But first I had to confine the little ones until they were big enough to fend for themselves against the Verraux Eagle Owls who also lived in my garden—this species of owl is the largest in Africa and the third largest in the world. I fed the little monkeys on carrots and peas and very gradually replaced these by leaves from the forest trees which would be their natural food when they were released. They were kept in a large wired-in enclosure in full view of Coli and his mother who often tried to reach them through the wire. When I eventually opened the enclosure doors the adults instantly adopted the little ones but although they always remained together, it took another six months before they really cuddled close together and started to groom each other. All four have lived happily together ever since, enjoying their freedom. I always have some carrots handy to give them, particularly the little ones as it helps to strengthen their teeth and bones.

The Hunting Hyena

Hyenas haunt many parts of the wild like so many sinister shadows. There are three species. The striped hyena is found mostly in India, Iran, Asia Minor, and North and East Africa. It's about the size of a wolf, greyish-brown in color, but with darker stripes across it. It has a mane along the neck and back, rather long forelegs, and, like all the hyenas, enormously strong jaws and teeth which enable it to crush the hardest of bones. The brown hyena of South Africa is about the same size: it's ashy brown in color, with a lighter collar, chest and belly. The spotted form, which ranges from Abyssinia to the Cape, is yellowish-brown, with darker spots. The striped ones, though, are the most numerous.

Of all the animals of the wild, hyenas arouse the most repulsion. This is partly because of their skulking habits—they have always had a reputation for cowardice, hanging about at the edges of other predators' kills, snatching at morsels when they have the chance, but slinking away at the first growl from the animal which has made the kill. It is also because of their fondness for carrion of all kinds—though they also perform a useful scavenging function, mopping up any offal left over from a kill. And above all, perhaps, it's because of their horrible cry, which has been compared to demonical laughter.

The hyena's reliance on carrion, though, was rather exaggerated in the past. In fact, it does a good deal of hunting. Its preferred prey is a gazelle fawn, though it frequently hunts adult gazelles as well. Ethologist Hans Kruuk has made some fascinating observations of the way the hyena hunts. It picks out an individual from among a herd of gazelles, and begins to chase it. An odd thing now happens. A gazelle is in fact a good deal faster than a hyena, but the hunted quarry gives the impression that it isn't really going all out. It keeps just ahead of its pursuer (at a speed of, say, 30 m.p.h.), at the same time leaping from the ground every now and then on all four legs, in a most spectacular manner. The purpose of this display, Hans Kruuk thinks, is to warn the other members of its herd. Then, when the hyena gets closer, the gazelle runs out from its herd, with the hyena at its heels. The chase continues for anywhere up to three miles. Usually the gazelle gets away, but in about one instance out of five, the gazelle tires and the hyena closes in for the kill. The prey makes a last minute attempt to escape by a series of sharp twists and turns, but by now it is too late: the hyena grabs it by the shoulder in its powerful teeth, then disembowels, dismembers and eats it, until hardly a scrap is left. Several other hyenas may join in the feasting, but the gazelle hunt itself is carried out by a single hyena.

It often used to be said that the hyena was usually a solitary hunter. As Hans Kruuk discovered, this isn't always the case, by any means, and he cites the example of the hunt of a wildebeest. In its first stages, it is true, this follows a similar pattern to that of a gazelle hunt. A single hyena will stand in the bright moonlight (it is a nocturnal animal) as a herd of wildebeest file past to or from their grazing ground. There's another odd feature here: the wildebeest hardly seem aware of the predator standing only a few yards away. Is it that, intent on their feeding, they just don't see or smell their enemy—or might it be that in some strange way they are geared to play their part in the whole ecological pattern of predator and prey?

When the herd has settled down to its grazing, the hyena will often walk close to it. The wildebeest go on grazing, grunting contentedly every now and then. Seconds before the hyena goes into action, a sudden silence descends; then the hyena dashes straight into the middle of the herd, scattering those in front, but letting them close ranks again. This apparently gives the hyena a chance to judge the paces of the animals —in much the same way, Hans Kruuk suggests, that a horse dealer studies the trot of a ring of horses for a good buy.

When the hyena has picked out a likely individual, he starts chasing it, and the wildebeest concerned dashes out of the herd and across the plain, entirely on its own. After a few restive moments, the rest of the herd settle down again to their grazing, almost, one might think, as if they had taken for granted that what had just happened was an inevitable law of nature.

Up to this point only one hyena, as a general rule, is involved. But as the chase goes on—for anything from two hundred yards to three miles—other hyenas join in, so that if the wildebeest fails to outdistance his pursuers, it will find, when eventually it turns to face them, that there are snapping jaws everywhere. It makes little serious effort now to defend itself, merely swaying its horns about in a rather aimless fashion, moaning, Hans Kruuk tells us, at the incessant ferocious bites. When it has been pulled down, as many as 52 hyenas may be feeding from the carcass.

The success-rate for the hyenas in hunting wildebeest is about one in three. In other words, horrible though the killings are, the various herds upon which the hyenas prey are not decimated: the proportion of survivors, presumably those most suited to perpetuate the species, is, generally speaking, ample to preserve the balance of Nature.

The calves of the wildebeest are born in January and February—and all of them in the course of a few weeks. This means that during this short period there are large numbers of the small beige animals among the black herds, sticking close to their mothers but obviously very vulnerable to the attacks of the hyenas, even though the calves can run quite well only an hour after they are born. At first sight this seems as if Nature has made a gross miscalculation. During these few weeks the hyenas, who display an uncanny ability to pick out the weaker runners, gorge themselves to the bursting point. But that, in fact, is probably the point. The hyenas just can't eat any more of the calves, no matter how greedy they are, and so some of the young wildebeests are bound to survive. The near-synchronization of births among the wildebeest calves, in consequence, serves Nature's purpose in the struggle for existence.

The prey which the hyenas like best of all, however, are zebras, presumably because a zebra carcass yields more meat than that of a wildebeest. And the zebra hunt is always a pack affair. The reason for this is that hyenas have a very healthy respect for the hooves and teeth of the zebra stallions which they do their best to avoid, concentrating on the mares and the foals—and obviously a single hyena wouldn't be able both to evade the stallion and attack his charges.

What usually happens is that a pack of hyenas (anything between ten and twenty-five in number) will advance on a family group of zebras at pasture —walking in that slow, steady, stiff-legged way that hyenas have—until they are about ten yards away. Suddenly aware of the presence of the hyenas, the zebras will raise their heads from their grazing. Both parties will then stand and look at each other for a while. Then the zebra stallion, his head low and menacing, will walk out from among his family (which might consist of four or five mares and perhaps a similar number of foals) towards the hyenas, and when he is a few yards away will start to gallop towards them.

Frequently that's the end of the business: the hyenas scatter (perhaps they sense that the stallion concerned is a particularly tough customer) and the stallion returns to his family. But in about one case out of three matters develop differently. The stallion will behave in exactly the same way, but this time the hyenas, while taking care to avoid him, will begin to spread out in a crescent round his family. Bunched together the mares and the foals will now slowly begin to walk away. At this point the stallion will charge at individual hyenas, attempting to bite and kick with his forelegs. But the hyenas, still steering clear of the stallion, will now spread out among the moving mares and foals, and pick an individual. Within a few seconds the doomed animal will be standing, strangely silent, among a seething mass of hyenas, while the stallion conducts the rest of his family to a fresh grazing place some distance away. Within an hour all that will be left of the prey will be a dark patch on the grass and perhaps the jaw-bone of the devoured zebra. Again there is a kind of fatalism on the part of the hunted herd and of the destined prey, almost as if they were taking part in an age-old formalized drama—as in a certain sense they are.

There's one thing about hyenas that doesn't seem to go at all with their reputation for cowardice—they will sometimes tackle a rhinoceros, that huge, armor-plated animal of the wild, which can weigh up to 3,000 pounds. As a rule, it's true, the attacks are more in the nature of a game, conducted by a party of hyenas in a boisterous mood—"rhino baiting." On one occasion Hans Kruuk watched as a mother rhinoceros and her small calf walked past a hyena den. Suddenly about twenty-five hyenas converged on the pair and tried to grab the calf. The mother rhino defended its calf ferociously, and the calf itself gave a very good account of itself. The battle went on for two and a half hours: at the end of that time the calf had lost its ears and its tail, and was bleeding from many wounds—but the pair were still holding out. Then another hyena appeared on the scene, hot on the heels of a fleeing wildebeest, and the other hyenas joined in that chase, rather than continue the more difficult task of getting the rhino calf. A fortnight later Hans Kruuk saw the calf again: it looked rather odd without ears or tail, but otherwise there seemed nothing wrong with it!

Wild Dogs: Solo in Danger

Hugo van Lawick

The word "dog" has such cozy domestic associations that sometimes it's difficult to remember that in several parts of the world the dog is also a wild species.

The African hunting dog is one of the wild species but looks quite different from either Australian dingoes or Asiatic wild dogs. They're distinguished by their rather peculiar color pattern—a mixture of yellow, black and white patches: rather like a tabby cat, in fact, and as with tabbies, the mixture of color is different for each individual. Another outstanding feature is the large rounded upstanding ears.

These African wild dogs prey mainly on the various species of antelope. The packs range over very large areas, in the search both for prey and for water, often abandoning their burrows on long and arduous treks in seasons of drought. Their home range, during the nomadic part of the year, might in fact extend to 1,500 square miles. They, too, in consequence seem characteristic denizens of the wild.

Wild dogs have always had a most unsavory reputation, because of their method of killing. Like wolves (and, if it comes to that, like domestic hunting dogs) they kill by disemboweling the living prey, though it is some small comfort to know that, as Hugo van Lawick and other observers have pointed out, the prey is in a state of psychological as well as physiological shock and so feels little pain. In fact, the wild dogs almost certainly dispatch their victims more quickly than the lion, which has been traditionally regarded as a "clean" killer. All the same, all those who have witnessed a pack of wild dogs pulling down a gazelle agree that it is a grisly sight. It might have been the sheer horror and ferocity of this hunt that led many observers in the past to assume that the pack acted in relays, impelled by a kind of communal blood lust, and that the identity of each dog was submerged in the identity of the pack, with no hierarchy or order of dominance, and no individuality.

Hugo van Lawick, however, was able to prove that, on the contrary, the social organization of a pack of wild dogs is as complex as that of a pack of wolves, with two entirely separate hierarchies for the males and the females. In 1970 van Lawick conducted the most detailed and enthralling study of an African wild dog pack that was ever undertaken. For a considerable period of time he trailed one of the dozen or so packs —which usually consist of between six and ten dogs— across the Serengeti Plains of Tanzania. For the general reader the most fascinating outcome of his researches was *Solo*, the story of a wild dog puppy and the pack to which she belonged, and one of the undoubted classics of wildlife in our times.

Solo was the daughter of a bitch named Angel, who had fallen foul of Havoc, the dominant female in the pack, and was in consequence ostracized and deprived of food by the rest of the pack. Solo was both the runt of Angel's litter and its only survivor, her brothers and sisters having been killed by Havoc. When the pack set off on a great trek across the Serengeti in search of food and water, Solo's chances of survival, in spite of her indomitable spirit and her determination to "belong," were slim. Here is part of Hugo van Lawick's description of Solo's attempt to keep up with the pack, while he and a friend followed in their car:

Several times again during the course of that long night a hyena almost caught Solo. Mostly one or other of the adult dogs noticed in time to race back and rescue the small pup, but on several occasions James and I protected Solo ourselves, moving the car between the predators and their intended prey. Normally we obey the rules, we observe nature as she is in all her beauty, her tenderness, her cruelty. Our task is to record, as accurately as we can, selected episodes from the endless story. But, rightly or wrongly, we had become involved with the life of this small scrap of dog. It had been so easy to view her actions through human spectacles and to admire traits in her which, had she been a member of our own species, we should unhesitatingly have labelled determination, resilience and, above all, pluck. And so, when her own kind failed her, we stepped in.

As the night wore on, however, we became increasingly certain that we were wrong to give Solo this protection. The pack, while it was less attentive to the small pup, was still aware of her. Every so often one of the adult dogs would turn back to wait for her, or carry her for a while. And so the progress of the whole pack was slowed down by our interference: but for us Solo would no longer be following and the dogs would have been free to hasten onward in search of fertile country where half-starved hyenas would no longer threaten the lives of all Havoc's pups.

The Gol Mountains were as barren as the plains. Wearily the dogs headed up one of the small foothills. They had travelled forty miles since leaving the den. When they had reached the brow of the hill Solo was left far behind, way down at the bottom. As the dogs paused, the faint calls of the lost, exhausted pup could be heard. Angel turned and stared back, then, very slowly, moved towards her daughter. The others watched. Angel called to Solo and, some ten minutes later, Solo appeared. She was tottering with exhaustion and when she stopped for a moment she swayed from side to side. When the pack moved on again Solo remained standing there, crying, unable to move. This time it was Rasputin who returned. He picked her up and carried her for a while. But probably he was tired too, and presently he put her down and she staggered along behind the pack.

It was clear that she could not continue much longer. The dogs had been circling the eastern slopes of the mountains, and now they looked out over the plains on the far side. There were a few trees here, amongst the rock piles. It would be hard to follow the dogs. Suddenly the pack quickened its pace: we chose to stay with Solo. She staggered another few steps, then

collapsed and lay down. She uttered a few calls of distress and was silent.

We were sure the other dogs had gone, and I got out of the car to examine Solo. But even as I did so there was a sudden explosive bark and, from the shadows, a dog came racing up. Solo had staggered to her feet at my approach, and now Yellow Peril ran up to her and stood, barking in threat at the car. He picked Solo up and moved on with her to where the pack waited.

Once more we followed, but almost at once Solo was left behind again. This time she did not even call: she just collapsed and lay still. She seemed a very small object in the moonlight. After a long time two shapes appeared, moving towards Solo through the shadows. We thought they were hyenas, come to claim their meal at last, but they moved off without ever noticing the small, still pup. Through binoculars we could see that they were two adult dogs, apparently searching for Solo. They moved out of sight, and did not reappear. Solo was very fast asleep. Possibly she was in the coma that precedes death.

From: *Solo, the Story of an African Wild Dog*, by Hugo van Lawick.

ᴏᴏᴏ

Hugo van Lawick and his companion decided to rescue Solo. Not without heart-searching, because they knew that to interfere with the course of Nature and to take the pup among humans might, in the long run, do her no service. However, they carried her back with them, and Jane Goodall, another distinguished naturalist who was then Hugo van Lawick's wife, took charge of the pup and gradually nursed her back to health. At the same time Jane Goodall and the others did their best to resist the impulse to make a pet of Solo, in the hope that eventually they would be able to return her to the wild.

When the time came to make the attempt they began to search for the pack to which Solo had once belonged—because obviously Solo would have a much better chance of readapting herself if she could do it through her own pack, in spite of the hard time it had given her when she was a weak and underfed pup. The search failed, but they did find Lotus and

Rinogo, a pair which had left the pack some six months previously when Lotus was in heat. Lotus now had a litter of five pups, and so the decision was made to try and introduce Solo to her, in the hope that Lotus would adopt her.

The van Lawicks had brought with them a portable cage, with a trap door which could be opened and shut from a distance. The idea was to put Solo in it and place it outside the burrow of Lotus and Rinogo. In that way Solo would be safe while the wild dogs investigated her and the van Lawicks observed their reactions. It wasn't easy to persuade Solo, who had never been confined in such a small space, to enter the cage, and in her fright she defecated—probably a fortunate occurrence because the smell would have helped to cancel out that of human beings.

So the cage was put in position near the burrow and the human watchers waited to see what would happen. Solo was frantic with excitement, but to begin with Lotus and Rinogo, who were lying outside their burrow with their pups, didn't notice the cage. Then suddenly Lotus heard Solo's cries, stared in the direction of the cage, and with a gruff bark of alarm jumped to her feet. This alarmed her pups too, and they beat a hasty retreat to their den. Rinogo now also got to his feet, and the two adult dogs moved, by cautious stages, towards the cage. When they got close they began sniffing, but almost immediately jumped back, Rinogo giving a sharp bark. Meanwhile inside the cage Solo herself was half mad with excitement and doing everything she could to show her pleasure and making all the usual gestures of submissiveness—grinning, flattening her ears, crouching low, rolling on her side, whining—and, of course, wagging her tail furiously. As she began scrabbling frantically at the floor of the cage, trying to get out, it was at least clear to the watchers that there was nothing she wanted more than to join her own kind. But there was still no way of knowing whether Lotus and Rinogo were impressed.

At this point the five pups came out of the den and also, very cautiously, began to approach the cage. Before they got there, though, Rinogo, as if he had decided to have nothing more to do with the affair of the strange pup inside her even stranger contraption, turned and walked slowly away. His pups followed him back towards the den. Lotus, however, stayed where she was for a while, examining the cage, sniffing at its occupant, and showing signs of concern. But

then she, too, turned and began to walk slowly back in the direction of the den.

The watchers were now in a dilemma. Solo was digging away at the floor of her cage more frantically than ever. But to let her out was perhaps to invite disaster. After all, Havoc had killed Solo's brothers and sisters who had actually been born within the pack —was it likely, then, that Lotus would tolerate the advent of an entirely strange pup? The human watchers decided to take a chance. Jane Goodall slowly pulled the string attached to the trap door. The movement at first frightened Solo, and she recoiled to the back of the cage. Then suddenly she realized that she could get out—and raced after the retreating Lotus.

The watchers held their breath. Solo reached Lotus, her tail wagging so frantically that it seemed to shake her whole body, leaped up at Lotus's face, and then flung herself on her back, grinning up at her. To the intense relief of the humans watching from a distance, Lotus, after staring down at the small creature at her feet, dipped her head and gave Solo's face a quick lick.

At that Solo leaped up and, her body still convulsed by the wagging of her tail, licked Lotus's face and raced on towards the den. Just before she reached it, the five other pups poked their heads out. Solo checked, and then dashed into another empty burrow near by. The other pups ran over to investigate. Two of them went down the burrow, and a moment later came out, accompanied by Solo. The other three pups now came up to her and began sniffing her. One of them jumped at her, whether playfully or aggressively it was impossible to tell. But it had the effect of causing Solo to race, *not* towards the empty burrow she had just left, but straight into the nursery burrow itself. The other pups followed her, and stayed there. Lotus stretched out at the mouth of the burrow, beside Rinogo, who remained aloof from the whole proceeding. Later on, as it grew dusk, all six pups came out of the den and ran over to Lotus. She stood up for them to suckle her—and Solo joined in as if she had been doing it every day of her life.

The van Lawicks and the other watchers were astonished as well as relieved. It was in their opinion probably the first time ever that a wild dog pack had experienced anything of this nature, and yet they had adopted the stranger with the minimum of fuss!

Birds in the Wild ... or, the Wild in Birds

Birds have soared over the earth for a very long time indeed. The earliest known birds, in fact, lived some 140 million years ago! They were about the same size as modern magpies. They looked nothing like them, though, with their jaws full of sharp teeth and their lizard-like tails—long, jointed and bony. But they had wings with which to fly, and the wings had the characteristic feathers which are, of course, so important in bird flight.

From the moment man himself evolved the birds have always exercised a special fascination, perhaps because they seemed part of the earth yet not really of it, soaring away from it into another element, reaching up to the world of spirits or of gods or to the gates of Heaven.

Their very difference from earth-bound man and the other animals inspired awe. It's true that some of them could be tamed to hunt or to be household pets, and others could be domesticated for the farmyard. It's true, too, that all birds don't fly. Indeed, according to many specialists it is probable that, millions of years ago, all birds ran along the ground, and that the origins of flight were the long, flying leaps made by the more energetic types. Apparently it's likely, too, that these early birds practiced parachuting—hurling themselves from tree to tree, as some reptiles and tree squirrels still do—before there was true flight. The idea of a long arboreal apprenticeship may be borne out by the gripping arrangements of the toes of birds, and also by the active climbing of the young of a primitive species of South American tropical bird called the hoatzin, which uses its hooked wings as well as its feet in moving from branch to branch.

The most obvious example of the flightless bird is, of course, the ostrich, which is also the largest living bird. There used to be six geographical races of ostrich, all differing slightly in size, shape and color. The last Syrian ostrich was killed and eaten by Arabs during the Second World War. The Northern ostrich, which belongs to the area south of the Atlas Mountains from Upper Senegal and Niger across to the Sudan and central Ethiopia, is the biggest, standing as a rule about eight feet tall if you take into account the long neck and head, and weighing between 265 and 280 pounds, though heights of up to nine feet and weights up to 345 pounds have been recorded. In the mating season the cock ostrich is considered one of the most dangerous of all animals and there are quite a number of instances of humans being killed at such times by this bird's powerful kick—so powerful that it has been calculated as equal to the knockout punches of five heavyweight boxers delivered simultaneously. It is also absolutely fearless, and has been known to attack a freight train, head-on. Perhaps it's just as well it doesn't fly—imagine being dive-bombed by an angry cock ostrich!

Still, it is flight that exercises the special attraction, and the combination of size and airborne capability is impressive indeed. The heaviest of the carinates (that is, flying birds, as opposed to the ratites, or running birds) is the Kori bustard of East and South Africa, which normally weighs between 25 and 30 pounds, though bigger ones have been known. The British Museum has the head and neck of a Kori bustard which weighed 40 pounds. With these sort of weights, however, flight is cumbersome and requires a long take-off run.

The wandering albatross of Antarctica, the magnificent seabird which sailors regard with such superstitious awe (and which figures so dramatically in Coleridge's famous *Rime of the Ancient Mariner*) can also weigh well over 30 pounds, though much of this often consists of fatty deposits which are gradually shed during long flights. The wandering albatross also has the widest wingspan of any living bird, measuring up to 12 feet.

Perhaps, though, it's those birds which soar above the mountain tops which appeal most of all to earthbound mortals, and which seem to be wild and free in a very special sense. There's the condor of the Andes, for example, which still ranges from Venezuela down to Tierra del Fuego and Patagonia. It's the heaviest of all the birds of prey, weighing anything from 20 to 25 pounds—which is about the weight of a good plump Christmas turkey. It certainly doesn't look like a turkey, though, when it is soaring, sailing and swooping among its native mountains, for it is one of the real monarchs of the wild.

Just as impressive, and better known of course, are the eagles. They, too, can be enormous birds both in weight and wingspan. The heaviest of them all is the harpy eagle, which ranges from southern Mexico to eastern Bolivia, southern Brazil and northern Argentina. The average weight is about 17 pounds, but weights of 20 pounds are not uncommon, and one specimen weighed no less than 27 pounds. But although these giants of the air arouse the deepest admiration, *all* birds fascinate man—including the tiniest of them all, the Bee hummingbird of Cuba and the Isles of Pines, which is also known as "Helena's hummingbird" or the "fairy hummer." This delectable little creature usually measures about 2.28 inches from beak-tip to tail-tip—and most of that *is* beak and tail—and weighs about 0.070 ounces. When you see it hovering and vibrating like a miniature helicopter among the tropical blossoms it is easy to mistake it at first for a rather large bumblebee or moth.

The most extraordinary thing of all about birds is that they can live in close proximity to man, sometimes indeed almost on top of him, and yet still remain inalienably in their own wild element. It is as if they carry their own special portion of the wild around with them. So near and yet so far from man, they are like the segment of another dimension intersecting our own. And this, above all perhaps, is the secret of their spell.

Birds Everywhere!

All birds are still subject to patterns and rhythms of behavior inculcated long before man made his appearance on the earth. It is the very stability of these patterns and rhythms that appeal to us today, because of the contrast it affords to the continuous restlessness and change of modern human life. It is both a comfort and a reassurance, reminding us that there may be parts of our nature, too, that obey immutable natural laws that are perhaps aspects of eternity itself.

Of all the age-old instincts implanted in birds those controlling migration probably strike us as the most spectacular. In this connection BBC broadcaster Robert Dougall had a remarkable experience one afternoon in September 1965. He had been working in his garden and after lunch settled down in his deck chair for a doze, in spite of the fact that the weather was stormy and unsettled. Suddenly he opened his eyes to find that his garden was alive with birds. What was more, he realized that there was an incredible variety of them. Scores of redstarts, with their "fiery, quivering tails" (the name derives from the old English word "steort," meaning tail) were busy in the hedgerows or feeding on the ground. Then he noticed the distinctive black and white of pied flycatchers. A moment later he saw wheatears, with their blue-grey backs, their wings black on top and buff underneath, and their white rumps—and he gives us the fascinating item of information that the shepherds of the South Downs used to call them "white arses," and that the name "wheatears" may well be a polite derivation from this crude country term.

The extraordinary thing about the appearance of the wheatears was that only a few of them normally make their appearance on this part of the English coast in early spring and summer. What could have happened to bring them to this peaceful garden in such unprecedented numbers?

Then, as Robert Dougall stared unbelievingly, he saw on the edge of a flowerbed a few yards away a bird which at first he couldn't identify. It was clearly exhausted, and its beige and grey mottled plumage was ruffled and bedraggled. It twisted its head in an oddly reptilian way, and Dougall realized it must be a wryneck, which is sometimes known as a snake-bird. Very cautiously he got up from his deck chair and looked around. He saw a number of other wrynecks, and in the hedgerows and the fields beyond he saw dozens of other exhausted birds.

By now he had realized that he was witnessing an astonishing ornithological phenomenon—it was, in fact, the largest fall of migrant birds ever recorded on the coasts of Britain. What had happened was that flocks of migrating birds on their way from Scandinavia and the far north of Europe to their winter quarters in southern Spain and Africa had been caught up in a great depression blowing up to the northwest from Italy. The birds had probably been struggling against it during the whole of the previous night and most of the following day, until, exhausted, they had thankfully found a landfall on this twenty-mile stretch of the Suffolk coast between Lowestoft and Minsmere, where Dougall has his home.

Later when he consulted his bird-watching colleagues in Minsmere he found that they had recorded no less than 52 species in that single astonishing afternoon, including wheatears, whinchats, redstarts, robins, garden-warblers, whitethroats, willow-warblers, spotted flycatchers, pied flycatchers, and tree-pipits. It was an extraordinary concourse of birds. Robert Dougall must have felt as if his garden had been invaded by another way of life!

There are many other instincts operating in birds, of course, besides the migratory ones (as other chapters of this book demonstrate), all of them forming a pattern of instinctual behavior full of subtleties and complexities, as well as of mysteries still to be solved. But at the same time they are fascinating to us because of their apparent immutability and self-sufficiency, in such marked contrast to the muddle and uncertainty —as it may seem to modern man—of human existence.

There is, for example, the innate "begging behavior" displayed by the chicks of the herring gull. The distinguished Oxford zoologist, Professor Niko Tinbergen, has described (in a classic study of the herring gull) how the chicks, a few hours after they are hatched, are begging to move about under the parent bird. This causes the parent to shift its position, and, sooner or later, to stand up and look down into the nest. Immediately the chicks begin to peck at the bill-tip of the parent bird they are in fact seeing for the first time, in the same instant usually spreading their tiny wings and making faint squeaking noises. These are irresistible signals to the parent bird. A large swelling appears at the base of the parent bird's neck and slowly travels upwards, producing all kinds of strange contortions, until suddenly it bends its head and regurgitates a lump of half-digested food. It drops this on the ground and detaches a small morsel from it for its chicks, which again dart their bills at that of the parent bird, not always accurately at first, but with perfect aim after three or four attempts. Sometimes they also peck at the lump of food on the ground, but for the most part they wait until another morsel, of a more suitable size, is presented to them. Then, when they have had their first meal, the pecking response dies away as quickly as it began and the chicks fall asleep. The parent will stand patiently with another morsel in its beak, probably calling once or twice, and then, when it receives no response, it will swallow the food again, afterwards carefully cleaning its bill either with its foot or by dipping it in the sand.

When the chicks are old enough to run about near the nest they display other innate behavior patterns. They react immediately to the alarm calls of their parents, stretching their necks and looking suspiciously around them; then, when the call is repeated, they run to shelter. There they adopt an inborn concealment process, crouching close to the ground and keeping absolutely motionless, so that their brownish-grey color will be practically indistinguishable from that of the surrounding sand.

In 1939 a scientist named Grohmann carried out an experiment with two groups of young pigeons to see if flying was learnt or innate. One group was reared in narrow earthenware tubes, in which it was impossible for them to raise their wings. The other group was allowed to go through, unhindered, all the "practice" movements characteristic of most fledglings. At the end of several weeks, the latter group began to fly, and the other group was then released from their confinement—and they too took wing! The performances of the two groups were carefully compared, but there was no difference in flying ability between them. In other words, the apparent "practicing" gave no advantage.

Survivors

The wild has not entirely disappeared, even in the modern industrialized twentieth century. It is, in fact, surprising how many of the species of wild animals which have been with us from time immemorial still hang on, in spite of the spread of housing, motorways and all kinds of pollution. An example which will spring to many people's minds is that of the wild pony. In fact, "semi-wild" would be the more correct description: the only true "wild" horse left in the world today is Przewalsky's Horse, named after the man who discovered it, at the end of the nineteenth century, in Mongolia. At that time there was one other species of horse surviving without help from humans, the Tarpan, but as it was then in danger of extinction it had to be taken under protection. A truly wild horse is one which not only lives wild itself but also one whose ancestors have never been tamed. The semi-wild ones, on the other hand, are usually the descendants of animals which once belonged to man but escaped.

A herd of wild white cattle still exists in Britain, grazing over moorland on the estate of the Earl of Tankerville at Chillingham, in Northumberland, as they have done for centuries past. They are probably the last survivors of the savage herds which once roamed all over Britain, and which are thought to be the descendants of the wild ox which inhabited northern Europe in prehistoric times.

This prehistoric ox was in fact almost certainly dark in color. The bulls in particular were nearly black, but apparently they had a white stripe down their backs. One of the theories advanced for explaining the whiteness of the British breed is that the Druids (the priestly caste among the ancient Britons—before the advent of the Romans), by a process of segregation and selective slaughter managed to produce a white animal for their sacrificial rites. What is certain is that the present herd is unvaryingly true to type: no colored calf is ever born into it. The cattle do have black muzzles and hooves, however (as well as black, or very dark, eyes), and when they are fully grown they have black tips to their horns and reddish hair inside their ears.

It's very likely that herds of these wild white cattle ranged the forests which once stretched along the border country between England and Scotland, from Chillingham to the Clyde estuary. Then·in the year 1250 one of the herds was rounded up and enclosed, either for food or for sport, within the wall which had just been built round the great park of Chillingham. When the reavers, or brigands, which at this time infested both sides of the border, made their frequent raids, they gradually eliminated the other herds, but were unable to get at the one inside Chillingham park—and there it remained and still remains well over 700 years later.

In all that time there has been no introduction of outside stock and no attempt at domestication. Indeed no one in his right senses would try and get close to these wild white cattle, which are almost certainly the most ferocious bovine species in the world. A fascinating fact about them is that they have preserved their instinct for preservation against animals which were once their natural enemies, but have long since disappeared from the scene. For example, they will ignore a shepherd's dog, but if a pack of foxhounds enters the park they will immediately form themselves into a close-packed group, as a preliminary to stampeding, if it should prove necessary. Then if the hounds do come close, the stampede will take place, in close formation, with the cows in front, the calves in the middle, and the bulls at the back, with the "king" bull acting as rearguard. There is, in fact, no danger from the foxhounds: it is the *pack* formation that triggers off the defensive mechanisms, because it is immediately reminiscent of a wolf pack.

The emergence of a "king" animal is, of course, as with so many other species in the wild, one of nature's methods for ensuring that only the best blood is passed on to succeeding generations. For it is only the strongest and fittest bull which becomes king, since in order to attain leadership (and the possession of the cows) he has to establish his superiority in combat over any other bull in the herd which challenges him. In due course he is himself inevitably beaten and another younger and stronger bull takes his place. Since fights to the death are rare among wild animals, the defeated bull's fate is usually a temporary banishment from the herd. Sometimes, though, he will come back and make a fresh bid for leadership—and sometimes he succeeds: one of the bulls in the Chillingham herd, in fact, reigned from 1947 to 1955, an unusually long spell. When an animal is sick, wounded, or simply very old, the normal savage—but necessary—law of the wild comes into operation: the animal concerned, knowing by instinct that it is now an outcast, normally leaves the herd of its own accord, and if it doesn't, then the herd quickly expels it.

The fact that no alien blood has ever been introduced into the herd throws important light onto the whole question of inbreeding in cattle, and the results in the Chillingham herd are being carefully studied in the hope that valuable lessons will emerge for the breeders of domestic cattle. The original size of the herd is not known for certain: a record of 1692 refers to "my Lord's Beastes," but gives no number. It is known, though, that during the nineteenth century the herd numbered between sixty and eighty. In 1925, however, it had been reduced to about forty, and after the particularly severe winter of 1947 all the young animals perished and only eight cows and five bulls were left. During the next eighteen months no calves at all were born, and it looked as if the herd was doomed to extinction. But then a slow process of recovery began: new calves were born, and by 1970 there were twenty-one females and thirteen males—and since then the gradual process of recovery has apparently continued. Incidentally, no matter what the size of the herd the ratio of about three females to two males seems to remain constant.

It would not be true to say that these cattle are wild in an absolute sense, for they have been protected insofar that for many years they have been neither hunted nor shot. Members of the public are allowed to visit them in Chillingham park at certain fixed times, so they have, in consequence, become accustomed to the sight of man. But they have retained their instinctive dislike of the human scent, and will not allow a near approach. They continue to breed without any outside assistance, so that besides being of great historical and scientific interest they really can be called survivors of the wild.

The magnificent wild red deer also still exists in Britain. This beast inhabits a few scattered places in northern England and as many as 150,000 live wild in Scotland.

The Death of a Bird

William H. Hudson

Finally, this beautiful description of the death of a bird provides a fitting epilogue to this chapter. It is by the writer and naturalist W. H. Hudson.

The bird, however hard the frost may be, flies briskly to his customary roosting-place, and, with beak tucked into his wing, falls asleep. He has no apprehensions; only the hot blood grows colder and colder, the pulse feebler as he sleeps, and at midnight, or in the early morning, he drops from his perch—dead.

Yesterday he lived, and moved, responsive to a thousand external influences, reflecting earth and sky in his small brilliant brain as in a looking-glass; also he had a various language, the inherited knowledge of his race, and the faculty of flight, by means of which he could shoot, meteor-like, across the sky, and pass swiftly from place to place; and with it such perfect control over all his organs, such marvellous certitude in all his motions, as to be able to drop himself plumb down from the tallest tree-top, or out of the void air, on to a slender spray, and scarcely cause its leaves to tremble. Now, on this morning, he lies stiff and motionless; if you were to take him up and drop him from your hand, he would fall to the ground like a stone or a lump of clay—so easy and swift is the passage from life to death in wild nature! But he was never miserable.

From: *Birds in Town and Village*, by W. H. Hudson.

ANIMAL VILLAINS

All wild creatures have their parts to play in the complex pattern of Nature, and in one way, of course, it is ridiculous to cast any of them as "villains." This is to apply purely human value judgments that cannot possibly be appropriate to animals. On the other hand, there are some creatures which, justifiably or not, have always inspired fear, terror or loathing among human beings; and there are some which, as their particular method of survival in the struggle for existence, have developed a degree of ferocity which, from a subjective human viewpoint, seems excessive.

However incorrect it may be scientifically to call them villainous, it is difficult not to do so. These are the creatures which often appear to kill, not out of necessity, but for the sheer love of it. And this is not confined to the mammals; among the birds, for instance, the shrike thoroughly deserves its popular name of butcher bird, as it apparently often attacks with its hooked, notched and toothed beak not in order to stock its larder with skewered victims, but out of an uncontrollable lust for killing—though here again it would be ridiculous to blame the shrike for doing what comes naturally to it.

Ursus Horribilis!

Earl of Dunraven

That's the Latin name for one of the biggest villains of them all, the grizzly bear. These bears belong mostly to Alaska, Canada, and parts of northwest America, though they are also to be found in the mountains of Europe and Asia Minor, across to the Himalayas. The really big American grizzlies which once inhabited the high peaks of the Sierra Nevada range, and which according to some authorities could attain a weight of 1,800 pounds, have mostly died out. But one which was killed in 1924, in the Okanogan Forest Reserve, Washington, and which, during the previous three years, was said to have slaughtered nearly 500 head of cattle and more than 150 sheep, weighed well over 1,000 pounds. In 1960, too, a sub-species of the grizzly, which had long been considered extinct, was re-discovered by oil survey teams some 150 miles north of Edmonton in Canada. It turned out that at least 400 of them were living in this remote wilderness, and they are now protected by the Canadian government. It is reliably reported that an average specimen measures 10 feet from the tip of the nose to the hind paw, and weighs about 1,000 pounds. There are exceptional individuals which are a good deal bigger than that, though it's worth bearing in mind that polar bears can be even bigger —one gigantic specimen weighed no less than 2,210 pounds and was 11 feet 2½ inches tall.

Grizzlies are big enough, certainly, to deserve the title Ursus horribilis. This account by the Earl of Dunraven, who travelled in North America in the 1870's when grizzlies were still abundant in the Rockies and other mountainous regions, bears out their infamous designation:

These bears behaved in a very singular manner. They scarcely ate any of the flesh, but took the greatest pains to prevent any other creature getting at it. I had hung a hind-quarter of one of the does on a branch, well out of reach, as I supposed, and had left the skin on the ground. To my great astonishment, on going to look for it in the morning, I found the meat had been thrown down by a bear, carried about 300 yards, and deposited under a tree. The brute had then returned, taken the skin, spread it carefully over the flesh, scraped up earth over the edges, patted it all down hard and smooth, and departed without eating a morsel. All the carcasses were treated in the same way, the joints being pulled asunder and buried under heaps of earth, sticks and stones. The beasts must have worked very hard, for the ground was all torn up and trampled by them, and stank horribly of bear. They did not appear to mind the proximity of camp in the least, or to take any notice of us or our tracks. A grizzly is an independent kind of beast, and has a good deal of don't-care-a-damnativeness about him. Except in spring, when hunger drives him to travel a good deal, he is very shy, secluded in his habits, and hard to find; very surly and ill-tempered when he is found, exceedingly tenacious of life, and most savage when wounded or attacked. Few hunters care to go after the grizzly, the usual answer being, "No, thank you; not any for me. I guess I ain't lost no bears," thereby implying that the speaker does not want to find any.

One day, while camped in the same place, Jack came in quite early, looking rather flustered, sat down, filled his pipe, and said, "J——s! I must have seen the biggest bear in the world. Give me a drink and I'll tell you." He then proceeded: "I was coming back to pick up an elk I had shot when I saw something moving, and dropped behind a tree. There, within sixty yards of me, was a grizzly as big as all outside. I watched that bear, quietly, but I'll be dog-goned if ever I saw such a comical devil in my life. He was as lively as a cow's tail in flytime, jumping round the carcass, covering it with mud, and plastering and patting it down with his feet, grumbling to himself all the time, as if he thought it a burning shame that elk did not cover themselves up when they died. When he had got it all fixed to his satisfaction, he would move off towards the cliff, and immediately two or three whisky-jacks, that had been perched on the trees looking on, would drop down on the carcass and begin picking and fluttering about. Before he had gone far the old bear would look round, and, seeing them interfering with his work would get real mad, and come lumbering back in a hell of a rage, drive off the birds, and pile up some more earth and mud. This sort of game went on for some time."

Smallest and Fiercest

But ferocity doesn't necessarily have anything to do with size. It is an astonishing fact, generally accepted among naturalists, that the fiercest mammal of them all is one of the tiniest—the common shrew! In some cases this little creature weighs no more than a seventh of an ounce. The fat-tailed variety of Europe is indeed the smallest mammal in the world, while the pygmy shrew of Virginia, Maryland and North Carolina, which is about three inches long, is the smallest mammal of North America. When these shrews are born, in a loose ball of leaves and grass usually tucked away in a hollow tree, stump or log, they are no bigger than bees. And even the so-called giant variety, the Asiatic water shrew, is never longer than nine inches—and a third of that is tail.

Weaned on flesh—usually earthworms—the young shrew fends for itself from about one month old, and its two-year life span is one ceaseless and savage hunt for food. Its minute eyes are practically sightless, and it sniffs out its prey or simply blunders on it in the grass, hurling itself upon it like a miniature tiger. Its preferred diet is beetles, slugs, centipedes, crickets, butterflies—insects of all kinds, in fact. The shrew begins shivering with excitement as soon as it finds its prey and gobbles it up in a few seconds. Its appetite is such that it can eat the equivalent of its own weight about every three hours, and it burns up energy at such a rate that if it is deprived of food it will die of starvation in less than a day.

Although insects form the bulk of its diet, it often kills and eats creatures twice its size. It is afraid of nothing. The American naturalist Alan Devoe has told how, as a boy, he put a shrew in a cage with a young white rat, intending to leave it there only a few moments while he got its own cage ready. But the shrew at once reared up, bared its teeth, gave a high-pitched chittering squeak of rage and hunger and hurled itself at the rat's throat. In no time every bit of the rat had disappeared, including its bones and claws and fur.

Another naturalist who kept snakes and fed them mice, once made the mistake of giving one of his reptiles a shrew to eat. When he went to the snake's cage next day, there was only the shrew left. It is in fact impossible even to put two shrews in a cage together, for in a few minutes the stronger will have eaten his weaker companion.

All common shrews have in their stomachs a gland secreting fluid with a very unpleasant odor which they can discharge when attacked by one of their few enemies. It has frequently been noticed that if a fox, for instance, pounces by mistake on a shrew, he usually drops it again in a hurry. The short-tailed shrew also has another protective device. Tests have shown that its salivary glands contain a poison similar to that of the cobra and other venomous snakes. When a shrew's teeth stab into an enemy, therefore, the victim becomes confused, finds breathing difficult, and is rapidly sticken by a form of paralysis.

Shrews do have their natural predators, such as great horned owls, weasels and bobcats, which can apparently stomach them in spite of the bad-smelling and poisonous secretions. Domestic cats can also deal with shrews, though they don't appear to eat them. One owner of three cats has related how frequently he comes down in the morning (cats, like shrews, are nocturnal hunters) to find two or three of the little corpses laid out for inspection on the kitchen floor. They look quite pretty, he writes, with their soft grey or fawn fur—except for the tiny hook-like upper incisors which are still bared in an expression of unrelenting and reckless ferocity.

Rats!

Rats are certainly a good deal bigger than shrews, but they can't be called large animals. Yet they have always inspired a fear and loathing out of all proportion to their size. In part this may be due to the fact that it was rats which in the past carried the dreaded bubonic plague which devastated whole centers of human population.

This species, in fact, was the black rat (*Rattus rattus*) which isn't anything as ferocious as the stronger and bigger brown or Norway rat (*Rattus norvegicus*) which has displaced the black rat in many parts of the world. If a black rat is attacked it usually turns and runs. If a Norway rat is poked with a stick or a broom, it will instantly leap at and bite its attacker. This variety is a burrowing rat. It can jump a few feet, and it can climb wherever there's a good footing. But its nimbler cousin, the black rat, can climb a tree, a drainpipe, a cable or even a concrete wall, as quickly as it can travel along the ground. It makes its nest in trees, under rock ledges or the eaves of houses or in lofts and attics.

Both species have voracious appetites, and will eat almost anything, including their own species if they have to. And both species multiply at a fantastic rate. The gestation period is only about twenty days, and they bear litters of four to ten baby rats four or five times a year—and those babies, blind and naked to begin with, are themselves able to breed at about six months old. It has been calculated that in theory a pair of healthy rats can produce a third of a billion offspring and descendants in only three years! It's this sense of gross fecundity, the feeling of the swarming millions, that largely accounts for the horror that rats inspire.

And of course they really *do* pose a serious threat to food supplies. They are the biggest pest with which the farmer has to contend. Harold Gunderson, a rat specialist at Iowa State College, has worked out the following rule of thumb for farmers:

If you never see a rat but do occasionally notice droppings or signs of rat damage, there are from one to 100. If you see them occasionally at night but never in the daytime, there are from 100 to 500. If you see many rats at night and several during the day, there are probably from 1,000 to 5,000 of them, and that figure is nothing unusual.

One of America's biggest poultrymen has claimed that rats cost him $10,000 worth of feed a year. Another reports that rats have killed 1,500 baby chicks in a night, and have carried off 80 dozen eggs a week. It has been calculated that every year rats eat or spoil as much food as 265,000 average farms can produce. Granaries, warehouses, the holds of ships and so on afford little protection. "Very few building materials can be considered absolutely ratproof," scientists in Savannah have reported. "Rats readily penetrate wood, have gnawed through most building-board materials and can in time penetrate heavy-gauge, hard-tempered aluminum." To test building materials against the ravages of rats, a cage was divided by a sliding panel made of plywood with a hole in one corner. Then each night a solid panel of the material to be tested was substituted for the plywood. By then the rats had got so used to going through the hole to get food that at night they gnawed at the spot in the test panel, no matter how hard and solid, which corresponded to the hole in the plywood one. In one of the Savannah tests it took rats just eight nights to tunnel through a two-inch panel of foamglass, while a panel of ordinary aluminum, half an inch thick, was penetrated by the rats in six nights. Of ten different grades of aluminum alloys tested, rats succeeded in drilling through all but one.

Man's fear of rats is increased not only by this extraordinary patience, persistency and adaptability, but also by the rodent's cunning. Some of this fear is irrational: there are all sorts of far-fetched stories about swarms of the animals being led by a King Rat of almost superhuman intelligence who can outwit any stratagem devised by ordinary humans. However, there *are* rat leaders, and a policeman on night duty in an English market town with docks and warehouses on the river reported that he often saw a column of rats, with an outsize one at its head, making its way through streets in the early hours of the morning, heading in the direction of the docks.

There are also plenty of instances of rat intelligence. Dr. Clarence M. Tarzwell, working at the U.S. Public Health Service laboratory near Savannah, put a cage full of freshly captured Norway rats into a pickup truck. Somehow one got out, and he found her more than a month later, still in the truck. She had made a comfortable home in the back of the seat and was raising a litter of nine thriving young. Meanwhile she must have got in and out of the truck dozens of times in search for food.

Another authentic story is of a caged female laboratory rat which planned an escape as shrewd as any devised by a human prisoner. Each night she gnawed at the wooden wall around the copper pipe which brought drinking water into her cage. Each morning, before the attendant came in, she skilfully covered the evidence with litter. The attendant had no suspicion of what she was doing—until it was too late. The same kind of cunning is often shown in evading the efforts of the exterminators. In any large rat community there are nearly always a few superior individuals who have learned not to nibble at poison and not to walk into traps. Exterminators have reported quite a few battle-scarred old rats which kick at a spring trap until it goes off, eat the bait, and then go off in search of another free meal. Even the most up-to-date chemical methods of extermination, applied in ideal conditions, expect no more than a 95 per cent kill—and the remaining five per cent of rats multiply so rapidly that within nine months to a year they are as numerous as ever. To make it worse, with this rate of increase, strains can develop which are immune even to the most sophisticated of poisons. The only worthwhile campaign against the rat, it has been said, is one that never stops—and that in itself is a pretty horrifying thought.

Glutton—or Devil?

Did you know that there was an animal named a "glutton" (*Gulo gulo* to give it its Latin name)? Its other name is "wolverine," and although it's another comparatively small animal—weighing between 25 and 30 pounds, a little bigger than an average terrier —it has a fearsome reputation. Its gluttony certainly plays a part in it: its appetite for the small mammals, birds and carrion which form its staple diet is truly amazing, and it's a real scourge to the trappers of the arctic regions of North America, where it is most abundant, because it is continually robbing their traps of both bait and captives.

At first sight it looks quite a cuddly creature, with its short legs, thick, bushy tail, and shaggy blackish-brown fur, with a broad band of chestnut on the sides of the body. But the short legs have large feet armed with sharp, curved claws, its teeth are like razors, and it's so strong that it can shift a log that would need two men to lift. And its greed is more than matched by its ferocity. It's afraid of nothing, no matter how big, and when it attacks, grunting and growling, it is always a case of kill or be killed. All the other animals of its habitat are in awe of it. Cases have been reported of cougars, and even grizzly bears, giving up freshly slaughtered prey to it, and of whole packs of wolves retreating before a single woverine. On one occasion a captured wolverine put into a zoo actually attacked and killed a polar bear. Two American naturalists have also reported an authenticated instance of a single wolverine attacking, and after a running fight, killing

a three-year-old bull moose. The female wolverine is especially ferocious when nursing her young, and if disturbed will think nothing of attacking an armed man.

The American Indians of the far north believe that the wolverine is an incarnation of the devil, not only because of its gluttony, strength and ferocity but, even more, because of its cunning. For some reason this seems to be particularly directed against man. The wolverine is one of the most inquisitive of animals, and, like the magpie, it delights in stealing and hiding all sorts of articles—including the trapper's gear. One of the reasons for the Indian's awe of the animal is that it displays an uncanny understanding not only of traps of all kinds, including the most complicated steel-spring ones, but also of every other kind of mechanical gadget, including guns. Often a wolverine, with its very keen sense of smell, will sniff out a whole line of traps during the night (it is a nocturnal animal) feeding on the trapped game until it is gorged full, and then tearing the rest of it to pieces just for the fun of it. The extraordinary thing is that wolverines coming into contact with men for the first time seem instinctively to understand their ways as well as those who have had previous experience of them.

The American naturalist Leslie T. White tells a remarkable story about a duel between a Cree Indian and a wolverine. During the first part of the trapping season the Indian had been very successful, and one end of his log cabin was stacked with the pelts of marten, ermine and fox. Then one morning he found that the whole of his trap-line had been destroyed by a wolverine. So he set out with his dog, a powerful wolf-husky, to track down the marauder. Eventually the husky flushed the wolverine into a thicket. The Indian rushed in after him and found the two animals locked together in a terrific fight. The wolverine had the dog by the throat. The Indian couldn't fire his rifle for fear of killing the dog, so he struck at the wolverine with the butt end. The wolverine released its grip and dashed off into the bush before the Indian could get in a shot. The husky's jugular vein was severed and it died shortly after.

The Indian now decided to build, at some distance from his cabin, a series of heavy log traps of the kind traditionally used by the Cree Indians and reputedly much more effective in outwitting the wolverine than the modern steel-spring trap. A blizzard came on and he was unable to return to his hut, so he made a camp for the night, hanging his snowshoes on the branch of a tree so that they would be safe from prowling animals. When he woke up the next morning, however, he found that the wolverine had somehow managed to pull the snowshoes down, and had then chewed through the frames and bitten the buckskin lacings into fragments. In that part of the country snowshoes are essential, so the Indian hid his equipment and set out to search for a willow tree, the branches of which he could form into a pair of makeshift snowshoes. When he got back to camp, less than an hour later, he found that the wolverine had torn his blanket into shreds, and had dragged away his rifle. Grimly the Indian sat down to make the snowshoes, and eventually, towards dusk, he was able to make the return journey to his cabin.

When he got inside he found that the wolverine had broken in in his absence, eaten or taken away everything edible that wasn't in tins, and destroyed every one of the precious pelts.

No wonder one of the names for the wolverine is glutton, and no wonder that the Indians regard him as the devil incarnate!

Driver and Army Ants

There are many other creatures which inspire terror, even though they are far smaller than any of the mammals. The smallest of them all are various species of ants. The most famous of these—and the most productive of horror stories—are the driver ants of Africa and the army ants of South America. One especially chilling story came from central Brazil in 1973. A column of army ants a mile long and half a mile wide was reported to have marched on the town of Goiania and devoured several people, including the chief of police, as well as numerous wild and domestic animals, before being driven back into the jungle by soldiers armed with flamethrowers.

This story has almost certainly been embellished by journalists' license, but both the African and the South American ants *are* pretty terrifying. When they are searching for food they march with extraordinary military precision, in parallel columns (usually four of them) six to eight ants abreast and flanked by guards, numbering in all anywhere up to 150,000 individuals. As they are blind, they proceed entirely by their sense of smell, and any living thing foolish enough to remain in their path is usually eaten. Although there are only a few authentic cases of human beings meeting this fate there are plenty of instances of animals failing to escape, especially tethered horses and cattle, or any other animal which is trapped or incapacitated in some way. On one occasion army ants in South America consumed a gorged, and therefore comatose, python. Usually the ants overwhelm their prey by sheer weight of numbers, and often they choke and blind it by crawling into its nostrils, mouth, throat and eyes. There's not the slightest doubt that any helpless human being left in their path—a baby or a wounded man for instance—would be doomed.

As a rule animals take care to get out of the way as soon as they sense that an ant army is on the march. Even the very formidable giant anteater of South America, which can mop up thousands of ants at a time with its long, sticky tongue, runs away from them. One eyewitness relates how, when driver ants invaded his house in East Africa, he saw beetles and cockroaches falling from every nook and cranny, covered with biting ants, and heard the rats in the rafters squealing in terror as they were attacked. He himself had great difficulty in preventing them from crawling up his legs and he had to spend the night outside the house. In the morning he found that the ants had moved on, leaving the building absolutely free of insects and vermin. In the garden he found that they had killed a young pet crocodile, though they had not succeeded in eating it.

While South American ants bite as soon as they encounter flesh, African driver ants crawl up the legs of quadrupeds and humans and only when they have gone some way (waist-high in a human) do they all bite in unison. It is generally agreed that the sudden shock of these innumerable and extremely painful bites could easily cause a child, or a sick or elderly person, to die. Moreover, the ants are very difficult to remove. Elephants are often goaded into a dangerous state of panic and fury when driver ants crawl up their trunks.

The author of that lurid report from Brazil was certainly right in one respect. Fire is the only way to stop a column of driver or army ants. The usual method is to soak the ground ahead of the column with gas and then set it alight. The fire must be kept going, though, otherwise the ants are likely to go on pouring into it—"like a funnel of treacle" one observer has said—until they have smothered it.

As far as the actual bite is concerned, though, the most dangerous ant in the world is the black bulldog ant (*Myrmecia forficata*), which belongs to the coastal regions of Australia and Tasmania. In attacking it uses its sting and its jaws simultaneously, so that "bite" is not really the right word. It is known for a certainty that this ant has caused several deaths. The most recently recorded was that of a woman at Launceston, Tasmania in 1963. She was "bitten" on the foot by black bulldog ants while in her garden, and died fifteen minutes later.

Killer Bees

From Brazil recently there have come reports of a new strain of bees which was giving rise to all kinds of horrific rumors. About seven years previously Professor Warwick Kerr, an American working at the University of São Paulo, had tried out an experiment aimed at improving the output of Brazilian honey. He crossed native bees with a strain from Tanzania, which is notorious for its aggressive qualities but which is also extremely energetic and produces excellent honey. To begin with, the new hybrids had proved a great success. But then their personality appeared to undergo a change. They had taken to attacking native bees and driving them from their hives. Then they had turned their attention to farm animals and dogs—and there were unconfirmed reports of human beings being stung to death by them.

There's no doubt at all that the new strain of bees was—and still is—a serious menace. And not only in Brazil. Before long swarms of them were invading neighboring countries. In 1974 they crossed the border into Guyana, taking two years to reach the densely populated coastal areas. One swarm descended on a woman who was cycling in the bauxite town of Linden, sixty miles from Georgetown, the capital. Luckily for her she was wearing a wig over her own hair. She pulled off the bee-covered wig, dropped her bicycle and dashed for safety. A businessman in New Amsterdam, seventy miles east of the capital, wasn't so fortunate. When he was attacked by a swarm of the killer bees he dashed for shelter, but was so severely stung that he had to be treated in a hospital. Three other people in another part of the region suddenly found themselves surrounded by the bees (the swarms seem to arrive and depart with mysterious speed) which immediately began to attack them about the head and face. Screaming in pain and terror they were rescued by three cab drivers. In a village only seven miles from Georgetown, a man pulled out the drawer of a cupboard, and found a number of the bees nesting there. Very wisely he slammed the drawer shut, sealed the cupboard and moved in with relatives until the experts could deal with the situation.

There have been no known fatalities among humans as a result of these attacks, but a number of people have been very severely stung and in several separate attacks donkeys and dogs have been stung to death. The attacks are so savage and the stings so painful that there's a distinct danger that sooner or later a child or an adult suffering from a weak heart *will* be killed. The authorities have been taking the threat very seriously indeed. Members of the public who come across a nest have been instructed to burn it, using a long pole to keep themselves at a safe distance. It is believed that these ferocious hybrids have the unique characteristic of mating only with virgin queens from other strains, and so the authorities are attempting to check the spread of the killer bees by destroying the virgin queens in the hives and replacing them with queens from a more domesticated strain that have already been mated. It has also been suggested that African birds which prey on the original Tanzanian strain Professor Kerr brought into Brazil for his experiments should be imported in the hope that they may also prove effective against this new hybrid.

There is every reason for treating the matter seriously. For one thing the killer bees apparently reproduce at the rate of 500 every week, and for another the swarms seem to move in the direction of the northeast trade winds. By March 1975 they had already spread to Surinam. By the end of the year they were attacking people in the towns, and the fire service had to be brought in to conduct a regular warfare against the bees, which nest in empty barrels, cases, abandoned cars and so on. The bees have now reached Argentina, French Guiana, and Venezuela, and as they are believed to advance at least 300 miles every year it is feared that they will soon reach other areas. It is known that they can cross expanses of water, and since the distance from Venezuela across the Gulf of Paria to Trinidad and Tobago is only some eight miles, they may before long have established themselves throughout the Caribbean.

Within the next few years they might also reach the U.S.A. via the Isthmus of Panama. Americans are very much aware of the danger. A team of scientists has been set up, headed by Dr. Otley Taylor of the University of Texas in collaboration with the appropriate authorities at the French rocket station at Kouru in French Guiana, in order to monitor the movements of the killer bees and to keep in constant touch with neighboring countries which are also frantically searching—so far without success—for means to exterminate them.

It really does look as if the bright idea which Professor Warwick Kerr had over twenty years ago has turned out to be one of those experiments, beloved of science fiction writers, which go disastrously wrong and raise all kinds of nightmare fantasies.

Monster Spiders

Henry Bates

For some people the ultimate in horror is the spider, and there certainly are some unpleasant ones. In fact all spiders (and there are 25,000 known species) are venomous, but fortunately only a few are dangerous to man. The most poisonous of them is probably the notorious "black widow" of the Americas, Hawaii and the West Indies. Only the female is dangerous because the male of the species cannot inject a lethal dose into a creature as big as a man. The black widow only bites when it's frightened, but even so it has been responsible for quite a number of deaths. Just as dangerous is the funnel-web spider of Australia (sometimes erroneously called a tarantula), which has also killed quite a few people over the years. The most dangerous of all the spiders, though, is the *Aranha armedeira* of South America, not only because it is the most venomous (and has the largest venom glands), but also because it is extremely aggressive. It has a liking for human habitations, where it often hides in clothing and shoes—and if it's disturbed it will immediately attack, biting not once but several times. As for the tarantulas which once had a fearsome reputation, not only in America but in many parts of Europe as well, they are apparently nowhere near as dangerous as has been claimed in some of the wilder stories—though it's certainly no joke to be bitten by one.

Some of the species of scorpions are much more to be feared than spiders. One type in North Africa can deliver a massive dose of neurotoxic venom (so called because the venom attacks the junction between the nerves and the muscles they control, causing excruciating pain) which has been known to cause a man's death within four hours, and a dog's in seven minutes. In September 1938 no less than 72 cases of dangerous stings from scorpions, mostly of this species, were reported in a single day near Cairo, though not all of them were fatal. Two years previously while nine men were sleeping in a guest-house on the northwest frontier of Pakistan, near Peshawar, they were stung by a scorpion, and eight of them died within a few minutes. During the desert campaigns in the North African desert during World War II, a number of both allied and German soldiers were killed by the stings of another species of scorpion. Three other species common to Mexico were responsible, the Mexican State Statistical Department reported, for deaths ranging from a minimum of 1,588 in 1943 to a maximum of nearly 2,000 in 1946, though thereafter the figures fell somewhat, owing partly to wiser precautions and partly to improved methods of treatment.

Somehow, though, the idea of a scorpion does not send the same shudder of loathing through most people that a spider does—and those who have a special aversion to spiders will shudder indeed at this description by Henry Walter Bates:

At Cametá I chanced to verify a fact relating to the habits of a large hairy spider of the genus Mygale, in a manner worth recording. The individual was nearly two inches in length of body, but the legs expanded seven inches, and the entire body and legs were covered with coarse grey and reddish hairs. I was attracted by a movement of the monster on a tree-trunk; it was close beneath a deep crevice in the tree, across which was stretched a dense white web. The lower part of the web was broken, and two small birds, finches, were entangled in the pieces. One of them was quite dead, the other lay under the body of the spider, not yet dead, and was smeared with the filthy liquor or saliva exuded by the monster. I drove away the spider and took the birds, but the second one soon died.

The Mygales are quite common insects: some species make their cells under stones, others form artistical tunnels in the earth, and some build their dens in the thatch of houses. The natives call them crab-spiders. The hairs with which they are clothed come off when touched, and cause a peculiar and almost maddening irritation. The first specimen that I killed and prepared was handled incautiously, and I suffered terribly for three days afterwards. I think this is not owing to any poisonous quality residing in the hairs, but to their being short and hard, and thus getting into the fine creases of the skin. Some Mygales are of immense size. One day I saw the children belonging to an Indian family who collected for me with one of these monsters secured by a cord round its waist, by which they were leading it about the house as they would a dog.

From: *The Naturalist on the River Amazon*s, by Henry W. Bates.

Bats and Vampires

Even more people, perhaps, have a phobia about bats than about spiders, and they have traditionally been regarded with fear and loathing. This is partly, no doubt, because of the creature's unnerving swooping and dipping movements, which are positively terrifying when it swoops into one's face or hair. Its soundlessness to human ears and its association with caves, ruins and other spooky places also frighten humans. But one of the main reasons for this mammal's offensive nature must be because nearly all of the world's enormous variety of bats have heads and tails which look like hideous caricatures of other animals—horses, foxes, cats, and rats. Even worse, some of them look like tiny, grotesque·human beings: the females of these species even have breasts in exactly the same location as those of a human female.

But of course the bat which has given the whole order of bats (the Chiroptera) a bad name is the vampire, so beloved of the writers of horror stories.

There are three species of vampire bats, all of them belonging to South America. They certainly aren't pleasant to look at, with their big ears, rat-like heads, pug noses, cleft lower lips—and their short, fantastically sharp incisors. Contrary to popular belief, though, they don't *suck* blood. What they do is to slice off a wafer of flesh with these incisors, as deftly as a surgeon wielding a scalpel, and then when the blood wells up into the wound they *lap* it up as cat laps up a saucer of milk. They do this in absolute silence, though, and so painlessly that if the victim is asleep he rarely feels anything or wakes up. As far as humans are concerned, the vampire's preferred point of attack is in between the toes. When the naturalist Charles Waterton was in the jungles of Guyana he slept with his big toe deliberately sticking out of his hammock, because he wanted to study the creature's habits. To

his annoyance the vampire ignored his obligingly exposed big toe, and instead got inside the hammock of a companion sleeping nearby—who woke up to find his whole foot soaked in blood. Sometimes, however, the bat will go for the forehead, fingers, ear lobes, and even lips.

These bats have tube-like stomachs, perfectly adapted to a blood-only diet, but the stomachs are so small that they can't take in more than an ounce of blood at a time—and that's usually all the bat needs each night. An ounce doesn't sound like much, but it does add up to two large buckets of blood per year per bat. If you think of a cave containing about a thousand vampire bats (quite a normal figure) the total would come to 5,750 gallons of blood a year—and that's a pretty gruesome thought! This would also mean that the thousand bats would need fifteen gallons of blood every night from the vicinity of their cave.

Although a vampire has been known to drain a bird dry, the amount of blood it could take from cattle or from a human being is not likely to cause much harm in itself, although the case has been reported of a boy in Trinidad who was visited by no less than fourteen bats in a single night, and that certainly involved an appreciable loss of blood. Generally speaking, however, the danger from vampire bats in the South American subtropics lies not so much in the letting of blood as in the fact that they are often carriers of rabies, and can spread epidemics among the herds of cattle, and sometimes infect humans too.

All the same, none of the bats, not even the vampires, are the fearsome creatures of popular imagination—though it's likely to be a long time before we disassociate them from ghosts, witches, blood-drained corpses and Dracula.

Crocodiles

There are thirteen species of the crocodile—as distinct from the alligators and caimans, from whom they differ in several respects though they are roughly similar in appearance. The largest of them, the estuarine or saltwater crocodile, which is found in many parts of Asia and Australia, is also the largest reptile in the world. A length of 14 to 16 feet and a weight of 900–1,150 pounds is quite common, and one gigantic specimen, shot in the Norman River of Australia's Northern Territory in 1957, was 28 feet 4 inches in length, and weighed a staggering two tons. Even bigger dimensions have been claimed, and it's possible that in the days when these creatures were left unmolested by hunters they grew even bigger than their descendants today.

None of the crocodiles are to be treated lightly, but the estuarine species is undoubtedly the most dangerous to man. It is estimated that it kills well over 2,000 people every year, and the figure may be closer to 3,000. In 1975 about a hundred people were on a pleasure-boat trip on the Malili River in central Celebes, Indonesia. The overloaded boat sank in the river, which was infested by estuarine crocodiles. Some of the passengers managed to get to safety, but 42 of them were eaten by the monsters. The same species inhabits the Nile River in Egypt, where it kills at least 1,000 people every year, most of them women washing clothes in the shallows or children playing there. At one time the annual figure was over 3,000.

The appetite and capacity of the crocodile are amazing. In 1823 two European settlers killed an estuarine crocodile near Lake Taal, in the Philippines, after a struggle lasting more than six hours. Inside the 27-foot monster's stomach they found a horse, bitten into eight segments, in addition to 150 pounds of pebbles of assorted sizes. One of the most terrifying aspects of this species of crocodile (though all of them are pretty frightening) is its prodigious strength. Some idea of it can be gathered from the fact that in 1939 an estuarine crocodile in northern Australia mounted a riverbank and caught hold of a Suffolk dray horse which had recently been brought from England, and which itself weighed no less than a ton. Although this breed of horse can exert a pull of over two tons, the crocodile succeeded in dragging it into the river. In several instances, too, a fully-grown black rhinoceros has lost a similar tug-of-war with an estuarine crocodile.

OCELOT CUB

CHIMPANZEES

F

Little Scaled Killers

There's a ferocious denizen of the rivers of South America which has a more villainous reputation than any of the crocodile family—and that is the tiny piranha fish.

There is a funny cartoon in a magazine which gives the reader a macabre chuckle. It is in two halves. In the first a man is standing on a riverbank clad in bathing shorts and with an anticipatory smile on his face as he prepares to dive in; in the second, a skeleton is scrambling out on the opposite bank, while a shoal of piranha gleefully churn the water behind it.

In fact, as is nearly always the case with these horror stories, the depredations of the piranha fish, which frequent the more sluggish waters of the big rivers of South America, have been much exaggerated. For one thing they rarely attack unless their victim is wounded and they can smell blood. Harold Schultz of the Natural History department of the Museu Pailista in São Paulo, declared that in all his twenty years of travelling he had only come across seven people who had been injured by piranhas, and these only slightly.

All the same, piranhas aren't to be fooled with. One early morning in Manaus, Brazil, the motorboat which daily delivers milk, letters, and the occasional passenger to the villages on the innumerable nearby creeks and small tributaries of the river was making its rounds. Among the handful of passengers was a ferocious looking individual in a bush hat, wearing an outsize revolver in an outsize holster, strapped round an outsize waist. When the boat reached a section of the river on the outskirts of Manaus the passengers

became aware of a most disgusting stink. They were passing the main slaughterhouse, the man at the wheel said. It was, as usual, a very hot and sticky day and one passenger was trailing his hand in the water, which was at least cooler than the air. Suddenly the man with the revolver let out a yell and yanked the man's hand back over the gunwale. At first the passenger thought he was being attacked by some desperado, until the man with the gun hissed "piranhas," and then explained that the one place on the river where one must *not* take risks was near a slaughterhouse, where the smell of blood was only too likely to attract the little monsters. In fact, a shoal of them was seen a little later. They looked as harmless as goldfish, but they had teeth as sharp as razor-blades, and jaws so powerful that they could have snapped off a finger like a carrot.

The piranha shoals can number up to 1,000, and if the smell of blood is about—and sometimes merely if the water is being agitated—they will launch an attack at lightning speed against anything, no matter how big. There is a record of a capybara (the largest type of rodent), which weighed 100 pounds, being reduced to a skeleton in less than a minute, and another of a wounded caiman (the South American alligator) stripped clean of flesh in less than five minutes.

Luckily, only four of the sixteen species of piranhas are dangerous to man, and most of the stories of human beings meeting the same fate as the capybara and the caiman are based on hearsay. But Harold Schultz himself admitted that he had once nearly lost a toe to the piranhas, and when Nicholas Guppy, another South American traveller, recently visited Apoera on the Courantyne river in Guyana, he found that nearly everybody there had lost fingers, toes, or sizeable chunks of flesh as a result of bathing or washing clothes in the river—while one boy from a village a few miles away had most of a foot bitten off by the piranhas.

As a matter of fact, there is a saltwater species almost as unpleasant as the piranha. This is the bluefish, which is found in many tropical or semi-tropical waters. It moves with the same lightning speed as the piranhas and, like them, travels in large schools. The bluefish was described in 1871 by Professor Spencer F. Baird, of the U.S. Commission for Fish and Fisheries, as "an animated chopping machine whose sole business is to cut to pieces and destroy as many other fish as possible in a given time." The bluefish has a quite incredible appetite and capacity. It has been estimated that an adult specimen of average size —which weighs about 5 pounds—can get through nearly a *ton* of fish (its choice fish are mackerel and herring) in a year. It isn't just a matter of necessity, though: the bluefish must be the greatest glutton in creation, cramming it stomach to the bursting point, and then disgorging the contents so that it can start all over again. Captured specimens are often found to have up to forty fish inside them.

At the same time, it's difficult to be sure whether it's gluttony or sheer lust for killing that's the motivating factor (insofar as one can use such words to describe a creature which is, of course, merely responding to instinctual patterns). Frequently it will destroy at least ten times as many fish as it can eat, and when a school moves in to attack, it does so with unnerving ferocity, churning up the water with innumerable little jets of spray, throwing bits of slaughtered fish about, and leaving a trail of oil and blood.

There's no doubt that if a human being fell among a hungry school of bluefish he could, in theory at any rate, be reduced to a skeleton just as quickly by them as by piranhas—though I've never heard of such a case. In April 1976, however, when a school of bluefish pursuing mullet along the Florida Gold Coast went into one of their characteristic feeding frenzies, at least a dozen swimmers were injured, and a long section of beach had to be closed.

Snakes Alive!

If some people go in dread of spiders or almost any kind of creepy-crawly, and others of bats or things that, as the old Cornish prayer for protection puts it, "go bump in the night," there are many others for whom the real nightmare creatures are snakes. They don't usually stop to ask whether *all* snakes deserve their condemnation. In fact there are about 2,500 species, and only a small proportion of them—probably no more than 650—are venomous; and of these only a very small number which are really dangerous to man. All the same, taking the world as a whole, between 30,000 and 40,000 people die from snake bites *every year*—and that's excluding China, the Soviet Union and some other regions for which figures aren't available.

The legendary asp, with which Queen Cleopatra is said to have committed suicide after the defeat and death of Mark Antony, may have been an Egyptian cobra. Cobras have quite short teeth, but the venom they inject with them is very potent. A single cobra may have enough venom at any one time to kill from ten to fifteen human beings. To make it worse, this snake usually holds on to its victim with its teeth, chewing away at the wound and all the time injecting more and more of the venom.

The biggest of the cobras is the king cobra which frequents South China, the Philippines, Vietnam, and Burma, and can reach a length of eighteen feet, and is, moreover, the most aggressive of them all. It's by no means certain, though, that cobras are the most venomous snakes in the world. As far as actual potency of venom goes, the various species of sea snakes (about 300 of them) take the lead, though fortunately they are all quite small and peaceable by nature unless frightened or attacked. Among the land snakes the dubious honor of being the most dangerous to man probably belongs to the extremely prolific saw-scaled vipers, which are to be found in great numbers in Africa, the Middle East and the Indian sub-continent. Not only is their venom unusually toxic to man, but they are also among the most aggressive of the snakes. They are followed pretty closely by the kraits of southeast Asia and the Malay Archipelago—and especially by the Indian species, half of whose bites prove fatal to man *even after the administration of anti-venom serum*, though some specialists say that the kraits of Java are even more deadly. Almost as deadly are the tiger snake of south Australia and Tasmania, the taipan of north Australia, and the bushmaster of Central and South America.

There are plenty of other dangerous snakes in South America (as in many other parts of the world), and naturally it's mainly in these regions that research into the various types of venom and the antidotes for them is carried out. One of the biggest and most famous of these research facilities is the unique Butantã Institute in São Paulo, Brazil. The snakes in the Institute, some horribly active, others comatose, lie in heaps on the grass like monstrous droppings, or hang from the branches of the trees like some disgusting trailing fruit. They are housed in a series of enclosures, and inside each enclosure is a number of low dome-like structures with slits at ground-level, covered with heavy lids, and inside these scores of other snakes sleep and digest their unmentionable meals. A sickly-sweet smell hangs over the whole park.

Every now and then a keeper, wearing leather gaiters and armed with a pole with a prong at the end, will step over the wall into one of the enclosures. Some of the snakes pay no attention, some come hissing round his ankles, and others glide out of the slits of the small blockhouses as inquisitive as terriers. The keeper will lift the lid off one of these, plunge in his pole and deposit a seething mass on the grass. As an encore he thrusts his pole into the branches of one of the trees and pulls: a dozen snakes fall with a thud at his feet. Another keeper uses the toe of his boot to provoke the dry-leaf skirr of a rattlesnake and another prods at a long flat snake which time after time rears and strikes at his gaiters with incredible speed and ferocity.

The climax comes when a keeper seizes a snake with his bare hands and squeezes its neck until the jaws gape. With a pair of pincers he indicates the two main fangs (there is a third spare one), then very-carefully draws aside the pellicle of flesh, soft and pink like an orchid, and squeezes again: glistening drops of venom shoot into the test-tube which the keeper holds in his other hand. After this the huge scorpions and the giant, furry spiders scuttling obscenely up the walls of their cages are an anti-climax.

Most of these creatures are sent to the Institute by hunters and settlers in return for phials of serum, indispensable for survival in the interior. In some years, one of the officials tells us, as many as fifteen thousand snakes are sent in. They are badly needed, for even this total, he explains, would yield only about two litres of venom a year from some ten thousand extractions, and by the time it was dried this would have been reduced to a bare half-litre.

Then there are the really big snakes which kill their prey not by injecting them with poisonous venom, but by crushing them and swallowing them whole. They belong to two main families, the boas and the pythons. Both have a very long ancestry, and their anatomic features lend strong support to the theory that all snakes originally developed from reptiles with limbs. An interesting difference between them is that whereas nearly all the species of python lay eggs, like the vast majority of snakes, the boas give birth to live young.

The real monster among the boas is the anaconda of the Amazonian jungles. In the past some fantastic dimensions were claimed. The early Spanish settlers in South America often reported anacondas of 60 to 80 feet long—they called the creature "matatora,"

or "bull-killer." In our own century some even more extravagant claims have been made. Thus in 1948 soldiers in south-west Brazil insisted that they had destroyed a 115-foot-long specimen by machinegun fire, and a few years later a Brazilian army patrol near Amapá, on the borders with French Guiana, reported that they had killed one measuring 120 feet. During the clearing of the vast forests in Central Brazil, preparatory to the building of the new capital of Brasilia, there were several stories of tractors hurled aside when they collided with even bigger giants. There's not much doubt that all of these claims were exaggerated. But there's no doubt about the specimen shot in Brazil in 1962, measuring nearly 28 feet, 44 inches *round its body* at the thickest point, and weighing about 400 pounds. It is, in fact, generally agreed among the experts that the maximum length for an anaconda is in the region of 30 feet. And that's hefty enough for any nightmare! There can be few more awe-inspiring sights than that of one of these gigantic creatures in motion.

Both the large pythons and the anacondas are capable of swallowing an enormous bulk. In 1955, for instance, there was a case of an African rock python, 16 feet in length, swallowing an impala (a species of African antelope) weighing 130 pounds—in other words not much less than its own weight! But there are several reliable reports of pythons and anacondas ingesting even more than that, and it has been calculated that an African rock python of between 25 feet and 30 feet could accommodate well over 150 pounds. The only comforting aspect of this is that the monster will then probably fall asleep for months (up to a year, and sometimes even more than that) before needing another meal.

There have been all sorts of lurid tales of both pythons and anacondas attacking and swallowing human beings. In fact, there are no authenticated reports of this happening in the case of the anaconda, and only a few in that of the python. Usually children are the victims, because the width of an adult man's shoulders would actually be too much even for the python's remarkably elastic jaws. In 1927, though, a fourteen-year-old boy was swallowed by a python on one of the islands to the south of the Philippines, and at about the same time it is claimed that an adult woman met the same fate. Also in 1927, an adult Burmese of rather small stature was ingested by a python. He had been on a hunting expedition in the jungle with four companions when they were suddenly caught in a violent thunderstorm. They were separated, and the victim took shelter under a tree. When the storm stopped, and he did not rejoin his companions, they started searching for him. Eventually they found his hat and his shoes lying close to a gorged python. They killed it and cut it open. Inside they found the crushed body of their companion: he had been swallowed feet first.

Charles Waterton, the nineteenth-century explorer and one of the first genuine nature conservationists, had a number of characteristically bizarre adventures with large snakes—boa constrictors—in the jungles of Guyana.

One day he was sitting on the decaying steps of Edmonstone's plantation house, reading his pocket Horace, when an excited native accompanied by his small dog, came running up to report that the dog's barking had led him to a large snake. Equally excited, Waterton followed him, picking up another native on the way. "I was barefoot," Waterton laconically states, "with an old hat, and check shirt, and trousers with a pair of trouser straps to keep them up." The snake was lying under a fallen tree, about half a mile into the forest. When Waterton more cautiously approached, he found that it was a coulacanara, or boa constrictor. The two natives begged him to let them fetch a gun. His response was to take away their cutlasses, for fear they might attack the creature and damage it. He then approached it again, and taking out his penknife, cut away the creepers that covered the branches of the fallen tree so that he could see where the head of the snake lay, tucked among its thick coils. Waterton had with him his "lance"—an eight-foot pole with an old bayonet attached to one end. Armed with this, and with the two natives behind him, he advanced—and even he admits that his heart "beat quicker than usual." He plunged the lance at the boa constrictor's neck and pinned it to the ground. The huge snake woke with a loud hiss, which sent the small dog running, "howling as he went." Handing the lance to one of the natives and ordering him to keep up the pressure, Waterton dashed forward to "grapple with the snake, and to get hold of his tail before he could do any mischief." With the help of the other native, he succeeded. Then he calmly took off his trouser straps and tied up the snake's mouth with them. It was an expedient that could only have occurred to Waterton; he doesn't report what happened to his trousers.

The boa constrictor objected to this indignity, and renewed the struggle. Eventually they persuaded it to wrap itself round the shaft of the lance, and set out to carry it back to the house, Waterton in front holding the head under his arm, one of his attendants supporting the belly, and the other the tail. The weight of the boa constrictor was such that they had to take frequent rests. By the time they got back it was too dark for Waterton to proceed at once with his dissection. So he put the massive snake in a large bag, and kept it in his bedroom. He did not, he confesses, sleep very soundly. The snake thrashed about all night, sending the bag bumping about in every direction, and it kept up a continuous hissing.

In the morning, Waterton untied the bag, and while his helpers held the boa constrictor down, he killed it. When he had skinned it, the Watertonian idea occurred to him of putting his head inside the mouth. He found he could do so easily, "as the singular formation of the jaws admits of a wonderful extension."

The Many-Armed Monster

"Devil fish" is the other name for the octopus, and it sums up the kind of fear and repulsion it rouses in most people. It has been the source of horror stories for centuries. Pliny, the first-century Roman natural historian, wrote of a fearsome 700-pound octopus with a head as big as a barrel. Eighteenth-century French sailors came home with a story of an octopus attacking their schooner, wrapping its tentacles round the masts, and nearly dragging it under the surface. The great nineteenth-century writer Victor Hugo sent shudders down thousands of spines when he declared in his novel *Toilers of the Sea* that an octopus could swallow a human being as a man might swallow an oyster. "The tiger can only devour you," he wrote,

"the devil fish *inhales* you. He draws you to him, bound and helpless. To be eaten alive is more than terrible; to be *drunk* alive is inexpressible."

The old serials of the silent screen often included an episode in which a diver lay helpless in the clutches of a huge octopus, while the man's eyes goggled in terror behind the glass of his helmet. Most of these horror stories must be taken with a pinch of salt. It's very rare for an octopus to attack a human being without provocation, and as a matter of fact it can be a most amiable creature. It's easy to tame, and can be trained to take food from an attendant or visitor. Some octopuses will even pull your fingers open if you hold the tidbit too tightly. They're also quite intelligent. An octopus in the aquarium at Brighton, England, learned how to get out of its tank during the night, make its way along a wall to another tank which held some small fish, help itself to some of them, and then return to its own tank. The zoologist Dr. Paul Schiller, while conducting experiments into the intelligence of the big molluscs at the Yerkes Laboratories in Orange Park, Florida, trained an octopus to push the lid off a jar in order to get at a crab inside. Another octopus was presented with a crab to which a white card and an electric wire were attached. When the octopus touched the crab, it got an electric shock. Then it was fed a crab without card or wire and got no shock. After only three experiments the octopus learned to leave the crab with the white card alone.

Some species of octopus are so tiny that even a fully grown one could sit comfortably on your fingernail. Most of the hundred or so varieties, indeed, are no more than three feet across. In the Mediterranean, where most are to be found, it's only the rare specimen which grows tentacles more than seven feet in length. So Pliny was putting us on a bit!

At the same time it's believed that some real giants *do* exist in the depths of the Pacific Ocean: it was reported that the tentacles of an octopus found in the stomach of a whale were 50 feet long—and that would mean that the whole creature was more than 110 feet across!

Whatever the facts—and even if we've discovered how delicious octopus meat can be to eat—the idea of encountering one of the creatures in the sea remains a nightmarish one, and most people would sympathize with Sir Arthur Grimble's feelings when he was challenged to capture one.

At the time he was a young district officer in the Gilbert and Ellice Islands of the Pacific, and, being new to the job, was anxious to make a good impression. One day he was standing at the end of the jetty in Tarawa Lagoon when he saw two young islanders getting ready to hunt octopuses (the Pacific islanders also regard the meat as a delicacy). One of the young men took a deep breath, dived, and disappeared. Grimble asked the diver's companion what was happening—and learned to his horror that the first man was offering himself as bait: he was swimming underwater close to the reef in the hope that an octopus, lurking in its cranny, would flash out its tentacles, seize him, draw him close to the cranny, and start groping at him with its horny mouth. At that moment the second man dived in, and a few moments later reappeared, with his friend beside him—and clamped to the friend's chest was a big, juicy octopus! As he arched his back in the water, in order to expose the whole bulk of the octopus, the second man leaned forward and plunged his teeth into the bulging eyes of the creature. The octopus died almost instantly: the hunters strung it on a pole, climbed back on the jetty —and calmly prepared to repeat the performance!

Intrigued as well as shocked by this bizarre method of hunting, young Grimble asked what exactly took place when both partners were under the water. It was quite simple, he was told: the rescuer caught hold of the "bait's" shoulders, at arm's length, and gave a great kick. This had the effect of tearing the octopus away from its proximal suckers, so that it clamped itself even more fiercely to its human prey. At the same moment, the prey himself also gave a powerful kick, which quickly brought him, octopus attached, to the surface. After that, it was up to his partner.

Young Grimble soon had reason to regret his inquisitiveness. The two islanders were busy whispering together. Then they turned to him and invited *him* to act as bait! It was great fun, they assured him, and there was really nothing to it. There was only one thing he had to be careful about—his eyes. It was best to cover them with a hand as he came close to the octopus, just in case a sucker closed on one of his eyes and blinded him: the pain, they explained, might make him expel his breath and inhale sea water—and that would spoil his buoyancy, and so make it much much more difficult for his partner to get at the catch and kill it. Grimble turned pale. He longed to turn and run. But he knew that the Gilbertese despise cowardice above all things, and *he* was supposed to be a leader. So, after swallowing once or twice, he gave a sickly smile and nodded his head.

He took off his canvas shoes, poised at the end of the jetty and then, with his heart in his mouth, dived in. He swam underwater close to the reef, hoping against hope that no octopus would notice him. But a moment later he felt something slimy and tremendously strong whipped round his left forearm and the back of his neck, binding the two together. He remembered to clap his right hand over his eyes, in the same instant that another "something" slapped itself against his forehead and then began crawling down inside the back of his shirt. His first impulse was to remove his right hand from his eyes and tear at the tentacle, but by now the right arm too was pinioned to his ribs. He felt his head and shoulders being jerked in towards the reef. A mouth began to nuzzle just below his throat. All the nightmare dreads of his boyhood swept over him.

A moment later, to his infinite relief, he felt a quick, strong pull on his shoulders. The nightmare came again as he felt the tentacles tighten convulsively round him, but he knew he had been pulled away from the reef. He remembered to give a kick, and as he reached the surface, to turn on his back. He saw the octopus sticking out from his chest like some giant tumor. His mouth was smothered by some flabby moving horror: the suckers of the octopus felt like hot rings pulling at his skin. A few seconds later his partner bit into the eyes of the octopus, and the weight fell from his chest. But those few seconds seemed to him "like a century of nausea."

When they got ashore the islanders danced round him, laughing and cheering, and throwing the dead octopus from one to the other. Grimble then saw that the "monster" wasn't by any means a big specimen. All the same, as soon as he had the chance, he found a quiet corner, and threw up!

Deadly Fleets

If anything can be called poisonously beautiful, it is the fearful and fascinating sea creature—or floating jellyfish colony—called the Portuguese man-of-war, which frequents the Atlantic, Pacific and Indian Oceans. This name was supposedly given some three centuries ago by English sailors, who, sighting a fleet of them with their brilliantly colored crested floats glistening in the sea off the Portuguese coast, thought they looked like a miniature collection of Portuguese galleons in full sail. It is not reported whether or not the sailors in the story subsequently realized what a deadly sight they were looking on, but beauty really *is* only skin deep in the case of the Portuguese man-of-war, because beneath its vivid iridescent floating dome there is a terrible arsenal which brings instant death to the smaller fish and plankton which are its prey, and which can have very serious effects even on man. Skin divers who have accidentally tangled with a Portuguese man-of-war have had to be rushed to a hospital in a state of shock, struggling for breath and bearing the marks of its stings for months afterwards.

Trailing down from the underside of every man-of-war are deadly fishing tentacles, one of which is far longer than the rest and is sometimes as much as a yard in length. Once these tentacles have stung and paralyzed the prey, they contract and haul the kill up to the extraordinary and horrible-looking mass of feeding polyps. The man-of-war has no mouth as such, the polyps attaching themselves to the victim and secreting a powerful enzyme which breaks down the proteins in its body. The poison in the tentacles' stinging cells is so strong that even a man-of-war stranded on the sand is dangerous; and when the venom—a neurotoxin—has been extracted scientifically, frozen, and stored in the laboratory, it has been found to last for as long as six years without losing its strength.

The man-of-war's beautifully colored float is actually a bladder some eight to twelve inches across, filled with a buoyant gas. For man, at any rate, it can serve as a warning flag—so long as it is visible on the surface of the sea. Unfortunately, however, this dangerous creature does not have to float; it can deflate the bladder and sink out of sight, re-forming its own gas and surfacing again when it pleases.

There is just one breed of fish—the nomeus—which appears to be not only immune to the man-of-war's poison, but actually swims in and out among the tentacles unharmed, snatching bits of food for itself from the deadly fishing lines. It even seems to act as a sort of living bait for the floating colony, sometimes swimming in ever-widening circles round it, until some larger fish gives chase. The nomeus darts into what must look like a tangle of floating seaweed to the pursuing fish—which is immediately caught up by the lethal tentacles. As its dead body is hauled up to the waiting polyps, the little nomeus starts nibbling at it as it goes.

Although there is as yet no known antidote to a man-of-war's poison (though some Bahamans are reputed to use urine as a treatment) and horribly impervious though these floating killers seem to man, they do have one natural predator. This is the giant loggerhead sea turtle which can weigh up to 500 pounds with a heavily armored shell and a powerful beaked head. There have been reports of at least one loggerhead being seen, swimming strongly through a fleet of men-of-war, its eyes swollen from the stings, but its mouth full of tentacles, leaving behind it a wake of broken, disarmed men-of-war.

More Jaws

Jacques-Yves and Philippe Cousteau

If there is one denizen of the deep who perhaps deserves the title of villain more than any other, it is the shark—or at least, some of the sharks. The greatest living experts on them, both in knowledge and first-hand experience, are probably Jacques-Yves Cousteau and his son, Philippe. In 1970 they published a remarkable book in which they described some of their underwater adventures when, from their converted mine-sweeper, the Calypso, *(supported sometimes by other vessels), they studied the behavior of these fascinating—and sometimes terrifying—creatures. They also made a world-famous film about them. Here Philippe Cousteau describes the blue shark of the Indian Ocean:*

His entire form is fluid, weaving from side to side; his head moves slowly from left to right, right to left, timed to the rhythm of his motion through the water.

Only the eye is fixed, focused on me, circling within the orbit of the head, in order not to lose sight for a fraction of a second of his prey or, perhaps, of his enemy.

His skin is creased with a thousand silky furrows at every movement of his body, emphasizing each pattern of incredible muscle. The crystalline water has ceased to exist; he is there in the unbelievable purity of the void and nothing separates us any longer.

There is no threat, no movement of aggression. Only a sort of nonchalant suspicion is apparent in the movements and attitudes of the shark, and yet he generates fear. Amazed and startled, filled with apprehension, circling with movements as slow and silent as possible, I try to keep him constantly in front of me.

There is something of the miraculous in the suddenness of his appearance as well as in his infinite grace; the surface of the water is far above and its absence contributes to the magical quality of the moment. He turns once more, and the sphere he encompasses expands or contracts, in accordance with his own primitive impulses or the subtle changes of the current. His silent circling is a ballet governed by untraceable mechanisms. The blue tranquillity of his form surrounds me with the sensation of a web of murderous and yet beautiful force. I have the feeling that I have accompanied his circular voyage since the beginning of time. His configuration is perfect. Suddenly, the idea that he deserves killing comes to me like a shock and instantly shatters the spell. Murder is the real function of this ideal form, of this icy-blue camouflage, and of that enormous, powerful tail. The water has returned to my consciousness, I can feel it again, gentle and flowing between my fingers, solid against my palms. I am one hundred and ten feet below the surface, in the clear, deep water of the Indian Ocean. With thirty minutes of air remaining and a camera in my hand, I am far from being an easy prey. Our circling has, in fact, gone on for only a few seconds and already I can hear the irregular snorting of the engine in the surveillance craft above me.

The great blue shark continues his approach toward me in the unchanging manner which has been that of his race throughout its existence. He is really a superb animal, almost seven feet in length, and I know, since I have often seen them before, that his jaw is lined with seven rows of teeth, as finely honed as the sharpest razor. I have already begun to ascend slowly toward the surface, simulating a few movements of attack whenever his orbit brings him sufficiently close. He perceives the slightest pressure wave from my smallest movement, analyzes every change in acidity or in the vaguest of odors, and he will never allow himself to be surprised by an abrupt movement. He can swim at a speed of more than thirty knots and his attack would probably be impossible to parry. But he is still circling slowly around me, making use of the cautiousness that has protected his species since its first appearance on this planet more than one hundred million years ago. I know that the circles are growing inexorably smaller and that I will probably succeed in repelling his first attack, but I also know that this will not discourage him. Startled for a moment, he will resume the circle of hunger, his attacks will become more and more frequent, and in the end he will break through my feeble defense and his jaws will close on the first bite of my flesh. Drawn by invisible signals, other sharks of the open sea will appear, climbing from the lowest depths or slicing the surface with the knife of their dorsal fins. And then it will be the scramble for

the spoils, a frenzy of hunger, of bloody and irresistible strength and horror. For this is the way of the great sharks of the open sea.

I climb back into our surveillance boat, the *Zodiac*, after a last glance at that flawless silhouette and the great staring eye, already regretting the impression of unconquerable power and exalting confrontation, cursing my weakness and being grateful for my fear. I look at the others, companions in dives like this one; burned by the sun, wrinkled by the sea, they look at me and understand: there is a shark beneath us.

While he was filming some of the sharks, Philippe had this terrifying experience:

I had exhausted the magazine of one camera, so I telephoned to the surface and asked that another be sent down. When it arrived, loaded with four hundred feet of new film, I was preparing to adjust the F-stop and focus when I glimpsed a dark mass just at the limits of my vision—about a hundred and fifty feet away. Then I began to notice that the sharks, which had been coming up to us more and more readily for the past half hour, were now nervous and wary. At first, I did not understand what had happened, since I did not associate the vague form I had seen with this new attitude of the sharks surrounding me. This situation went on for several minutes, and then, at last, I understood everything. The cause of the sudden agitation was approaching.

I recognized it as one of the most formidable of the deep-sea sharks, a great *longimanus*, well known to my father and all of us, more than nine feet in length and accompanied by at least eighteen pilot fish, each of them a good-sized animal. It was because of this moving cloud that I had not at first recognized the shark in the somewhat cloudy water.

While the brute strength of other sharks is tempered by their beauty and the elegance of their form and movement, this species is absolutely hideous. His yellow-brown color is not uniform, but streaked with irregular markings resembling a bad job of military camouflage. His body is rounder than that of other sharks and the extremities of his enormous pectoral fins and his rounded dorsal fin look as if they had been dyed a dirty gray. He swims in a jerky, irregular manner, swinging his shortened, broad snout from side to side. His tiny eyes are hard and cruel-looking. The cloud of pilot fish changes shape, sometimes scattering and then drawing closer together in an uncertain, nervous rhythm. From time to time, one fish will detach itself from the group and go off to inspect an object of some kind, then hastily return and take up its former place. Two large remoras and one smaller one form dark spots on the shark's belly.

I had a vague consciousness of sudden silence, and when I became more fully aware of it I realized that I myself had been breathing more slowly, almost as though I was attempting to hide. A few feet in front of me, Marcel too had forgotten his work and was watching the intruder. The big white-finned shark had disappeared, and the others were swimming rapidly, furtively, keeping well away from the newcomer. Sharks accompanied by pilot fish have already been compared with great Flying Fortresses surrounded by a squadron of fighter planes, and it is an image which gives a clear impression of the destructive force embodied in this vision.

This particular shark was swimming in lazy circles, about 50 feet from our cages, but his mere presence had brought the scent of fear to this little corner of the ocean. A few more minutes passed, and then I reacted at last. I was not going to permit this animal to ruin our dive, so I signaled to Marcel to go on with his work. A morsel of fish was waved about outside the cage, and when a small shark darted forward to seize it, Marcel managed to mark him perfectly. While I was checking my camera to see how much film remained, I saw the *longimanus* again. He seemed to be paying no attention to us and had even moved a trifle farther away. Marcel had no more fish with which to attract the other sharks within range of his spear, so I used up the rest of my film on random shots of the sharks remaining in the area. I was on the point of signaling for us to be brought to the surface when I was suddenly surrounded by a rustling flight of black-and-white pilot fish. They had left the shark, as if at some mysterious signal, and were circling about me like a swarm of moths around a flame. About 50 feet from my cage, the great ocean shark turned suddenly and hurled himself forward at incredible speed. In a fraction of a second he was beneath the stern of the *Calypso* and had snatched at the gleaming casing of the transmitter-receiver of the undersea telephone, which was hanging just below the surface. The cable was sliced in two as cleanly as if by a giant pair of scissors. The shark turned violently back on himself and furiously coughed up the metal box, which promptly sank to the bottom. Without a moment's pause, the shark arrowed his huge body toward Marcel, who had just time enough to close the door of his cage. The shark ricocheted away like a bullet, and turned straight toward me, seizing the bars of the cage in his jaw—a scant six inches from my face—and shaking it like a madman. I had a vision of the rope to the surface being cut and the cage drifting down, leaving me no alternative but to try and

get back to the ship on my own, exposing myself to this frenzied attack. And then, abandoning the twisted bars of the cage, he turned again and disappeared as swiftly as he had attacked, followed by his straining escort of pilot fish.

It seemed to me that I remained there for an eternity, motionless, almost without breathing. I had not had time to be afraid before. Marcel was watching me, and I was dimly aware of a great plume of air bubbles around his head. At last, I could feel the cage going up, and I emerged in the blinding sunlight.

How dangerous are *sharks to man? In his summing-up, Jacques-Yves Cousteau examines this question. He divides his conclusions into two sections—one for optimists, and the other for pessimists. On the optimistic side he decides that the danger need not be all that great,* provided *all the precautions are taken and the human beings in contact with the sharks behave coolly, sensibly and with sufficient knowledge of the different species and of their characteristic behavior. In fact, his conclusion as far as genuine underwater work is concerned is: "All things considered, diving in tropical waters is actually much less dangerous than riding a motorcycle"—though it must be remembered that as the accident statistics show, that* is *a pretty dangerous occupation. And Cousteau is careful to give equal weight to the pessimistic side of the question:*

Every species of shark, even the most inoffensive, is anatomically a formidable source of potential danger. On paper, the most to be feared are the great white sharks (*Carcharodon carcharias*), with their enormous jaws and great, triangular teeth. But, in reality, this species is extremely rare. By far the most disturbing are the *Carcharhinus longimanus*, whose great rounded fins bear a large white circle at their extremities. These "lords of the long hands" are encountered not only in the open sea, but everywhere in warm waters. They are the only species of shark that is never frightened by the approach of the diver, and they are the most dangerous of all sharks.

The youngest sharks—and therefore the smallest— are the most brazen. Even a very small shark, two feet in length, can inflict dangerous wounds.

Sharks race in from great distances to devour any fish in trouble. They can perceive the fish's convulsive movements by the rhythm of the pressure waves carried to them through the water. At a short distance, sharks are also extremely sensitive to odors, and particularly to the odor of blood. For both these reasons, underwater fishermen should not attach their catch to their belts.

Sharks are accustomed to attacking, without fear, anything that floats. They may, therefore, hurl themselves at the propellers of an outboard motor. This attitude makes them dangerous to swimmers, especially if the swimmer splashes about a great deal and makes considerable noise. For divers, the moments of entering and leaving the water are particularly dangerous.

The smallest bite of a shark is very serious, and may perhaps be fatal, since it always involves a considerable portion of flesh. In addition to this, the effect of shock is proportional to the quantity of damaged flesh. A victim of shark attack may die as a result of shock, even if the part of the body damaged by the animal's teeth is not vital.

There still exists no effective means of keeping sharks away from the area in which you are diving—either by chemical products, by sound waves, or by fields of electricity.

It is dangerous to dive at night or in troubled waters, and especially if there should be sharks in view, without using some strong protective device, such as a solid antishark cage.

It is dangerous to show fear of a shark; he knows this by instinct, and can profit from it.

It is dangerous to unleash the defensive reactions of a shark by attacking him (with a spear, a rifle, an explosive, or an electric shock) or even by frightening him (by pursuing him into a place from which there is no escape, for example).

When sharks are gathered together in a group, their behavior is unpredictable. A "frenzy" may suddenly take place, for reasons of which we still know nothing.

From: *The Shark*, by Jacques-Yves Cousteau and Philippe Cousteau.

Killer Whale!

When the film industry cast round for a fitting follow-up to the horror film *Jaws*, they could hardly have chosen a better one than *Orca*. *Orcinus orca* (or *Grampus orca*), better known as the killer whale, is a more terrifying creature than any shark. It isn't all that big, as whales go—reaching a maximum of about 30 feet. But it is undoubtedly one of the most terrible predators on earth. For one thing, it's built for speed, manoeuverability, and power. The killer whale has a rounded snout, its body is hairless, like the porpoise, and tapers back so that it is perfectly streamlined. The two broad horizontal fins (or flukes) at the end of the tail, together with the fore flippers, help it to co-ordinate its body movements and to thrust through the water, and to twist and turn without slackening its speed for a moment. A porpoise can swim at 25 m.p.h., but a killer whale can reach 34 m.p.h., making it one of the fastest swimmers in the sea. In each jaw it has ten to fourteen pairs of heavy interlocking teeth, exceptionally strong and measuring about 3 inches in length and 2 inches in diameter, truly fearsome weapons of destruction. As it is also one of the most intelligent of the whales, it can easily be seen why every other creature of the sea, including not only giant sharks but other whales, even those much larger than itself, are afraid of Orca.

Its preferred diet is fish, but it also enjoys warm-blooded sea mammals, and even penguins—and on occasion it has been known to leap out of the sea, seize a low-flying bird and swallow it. Its appetite is enormous. It consumes the smaller fishes by the hundreds, and the stomach of one 21-foot specimen was found to contain 13 porpoises and 14 seals. Incidentally, it is believed that after swallowing a seal the whales flay it and then disgorge the skin.

Killer whales are awesomely ingenious and aggressive in their search for food. Naturalist William Cromie has seen them swim under icefloes on which seals were sunning themselves, tip the floes up with their huge backs, and so shovel the seals into their gaping jaws. He has also seen baby walruses butted off their mothers' backs, where they had taken refuge, and quickly gobbled up.

On another occasion Cromie and a friend were studying a herd of Weddell seals on the ice. One of the seals, a large one about 10 feet long and 7 feet round his middle, was sleeping near the edge of the ice, some distance from the rest of the herd. Suddenly the two men spotted the sabre-like fins of a number of killer whales slicing through the water about half a mile out. Cromie ran over to the sleeping seal, kicked it awake, shouted, and then turned and ran to safety. He thought that his action had saved the big seal's life, but when he looked back he saw that it was still sleepily staring round him, in exactly the same spot. A few seconds later a huge head shot vertically out of the water a few feet in front of the ice. The killer whale then hurled the front part of its body across the ice, sank its teeth into the seal's hide and dragged the 1,000-pound animal over the edge, "as if it were a stuffed toy."

Killer whales travel in packs of anything from three to forty. The males sometimes go ahead in smaller groups—and when one of these groups goes into concerted action it is a terrifying spectacle. In his book,

The Cruise of the Cachalot, Frank Bullen relates how, on a whaling voyage in the North Pacific, he spotted a bowhead—a slow, big-headed whale. The bowhead was obviously in trouble, but Bullen couldn't at first make out what was the matter. Then a killer whale leapt right out of the water and landed with its full weight on the bowhead's back. He repeated this time after time, as if trying to pound the bowhead into submission. For some reason the victim seemed unable to swim away or to fight off the leaping killer. Then it raised its enormous head out of the water—and Bullen saw why: two other killer whales were hanging on to the bowhead's huge lips, as if trying to drag its mouth open. While they hung on in this way the other killer kept up its merciless pounding until the bowhead was exhausted. Then this killer too went for the bowhead's lips. After a short struggle the three of them succeeded in prying the mouth open, and then devoured their victim's tongue. "This was their sole objective," Bullen wrote, "for as soon as they had finished their barbarous feast, they departed, leaving it helpless and dying." This horrific practice has been confirmed by other observers.

At one time it was thought that killer whales never attacked human beings. It's true that *Scott's Last Voyage*, which recounts the story of the ill-fated expedition to reach the south pole, led by Captain Robert Falcon Scott in 1911, records a combined assault on the party by a group of killer whales, but it was assumed that the attack was really directed at the dogs, probably mistaken by the whales for seals. Only a few years ago, however, during the British Terra Noval expedition, Herbert Ponting, the official photographer, was focusing his camera on a group of killer whales out in the bay when suddenly the three-feet-thick ice under his feet heaved up and cracked. There was a loud, blowing noise, followed by a blast of hot, acrid air smelling strongly of fish. Eight killer whales had come up under him, broken the ice beneath him, and cut him off on a small floe. Then this floe began to rock furiously to and fro as the whales shoved their huge heads out of the water. Ponting jumped to a nearby floe, then to another, and another . . . and the killer whales followed him, snapping at his feet, he explained afterwards, just like a pack of hungry wolves. By the time he reached the last of the small floes it had drifted too far from the solid ice for him to attempt the jump. Luckily for him the currents caught hold of the floe and drifted it back. So, still clutching his camera, Ponting made one last leap. As soon as his boots struck solid ice he started running. He looked back just once and saw "a huge tawny head pushing out of the water and resting on the ice, looking around with its little pig-eyes to see what had become of me."

So, perhaps *orca* is the most ferocious of all the animal villains. It mustn't be forgotten, though, that this word "villain" is really meaningless in the animal context. The killer whale, in fact, plays an important part in helping to control the balance of Nature in the oceans. As Dr. Carleton, of the New York Aquarium, has pointed out:

"The most rapacious predator on earth is the two-legged one, man. But man seldom serves Nature's purposes. As a fisherman, he kills indiscriminately. As a hunter, he 'takes off the top,' usually slaying the prime animal. In contrast, the natural predator takes off the bottom. The wolf takes the weakling, the stray. The killer whale does the same in the sea, leaving the best and strongest to survive and breed."

Of Beasts and Brutes

Elspeth Huxley

Finally, Elspeth Huxley, the well-known naturalist and traveller in Africa, challengingly reflects for us on the whole concept of villainy in animal creation:

So deeply is the urge to hunt and kill our dinners implanted in the human heart—or, to be more precise, in the human genes—that it takes more than the changed circumstances in which we all now live to subdue it. Our attitude towards the other animals with whom, unfortunately for them, we share this planet, is fixed not only in our genes but in our language, hence in our minds and our behavior. The words we use betray quite plainly that, towards most animals, the human stance is one of distrust mingled with contempt.

If we want to be really rude about each other—which, to judge from the abundance of insulting terms, we do—what do we compare each other with? An animal, of course. Don't be so beastly, we say: i.e. like a beast, horribly unkind. He's a brute. "One of the lower animals as distinguished from man" is the definition of a brute; we have twisted the word to mean a person who is cruel, coarse and thoroughly nasty. Then look at some of our commonest terms of abuse: swine, pig, hog, bitch, snake, rat, jackal, hyena. This contemptuous attitude colors our similes and metaphors. Lie like a stoat. (When, and how, do stoats lie?) Dumb as an ox. Crafty as a fox. Chicken-livered. Crocodile tears. Verbs derived from animal orders and families are derogatory: toad-ying, ape-ing, badgering, rat-ting. Rats leaving sinking ships. (Any rat who voluntarily remained would be crazy; and do not humans leave such ships?)

Such exceptions as there are mainly favor animals that man has domesticated and so turned into slaves. Hounds may be faithful, steeds gallant, lambs innocent and so on. Birds form the one class of animal that has, to a large extent, escaped this linguistic condemnation. Owls are wise, eagles swift, but even birds are not entirely exempt from human disapproval: you may have a voice like a parrot, or, if ruthless and rapacious, be likened to a vulture. When humans endeavor, as they very occasionally do, to be complimentary to an animal, they nearly always get it wrong. The classification of politicians into hawks and doves suggests that doves are gentle, peace-loving creatures coo-ing amicably to each other on a shared bough; in fact they are quarrelsome, bad-tempered, greedy animals, a good deal more aggressive towards their co-specifics than the solitary-dwelling hawk.

Most of us look down on some of our fellow-humans, wrong though we may know this to be; it takes an exceptionally self-assured individual to look down upon them *all*. That is where animals come in. We can, and do, look down upon the whole animal creation from the height of our position high up in the evolutionary tree. Animals are the great ego-boosters of the human race.

Because these attitudes are embedded in our language, which in turn shapes our thoughts, they go unchallenged, and until comparatively recently no one bothered to find out just how animals really do behave. We are now beginning to realize that their actual behavior is not only quite different, but the reverse, of all that we had supposed.

"He gave way to his animal instincts," says the judge of a man convicted of rape. Animals do not indulge in rape; the consent of the female is always sought and given before mating takes place. "Worse even than a beast," we say of a man who batters his baby to death. No beast ever batters his baby to death. Mothers may now and then swallow them whole, but that is an attempt, however ill-judged, to protect the helpless creatures from harm.

Now that the study of animal behavior has become a science, ethologists are returning from lengthy sojourns among the beasts with observations which contradict almost everything that had been previously taken for granted by most of mankind. Take, for example, gorillas. Ever since du Chaillu returned from the Congo with blood-curdling stories, totally false, about huge hairy gorillas who charged through the jungle beating their breasts and intent on crushing to death the helpless human, gorillas have been seen as fearful creatures, menacing to man. Then came Dr. George Schaller, who lived among gorillas for a year or more and found them to be gentle, non-aggressive, shy and even timid creatures, living in affectionate family groups in a most un-humanly harmonious way. The chest-beating and roaring is merely display, to make the point that the territory you are invading is his, not yours—an anti-trespassing demonstration. If you retreat, so does the gorilla; and even if you do not, the gorilla will not actually attack you.

Gorillas in their native state, devoted to their young, put up with a sex-life that would fill most human males with horror were such restraints imposed upon them. During the whole course of his study, Dr. Schaller saw only two copulations; he believed that a male gorilla may go a year or more without a single one—and with no signs of frustration, such as approaches to other males. The dominant male has his family circle of wives and children, but is tolerant and not, apparently, jealous, when one of his wives enjoys a flirtation with a young bachelor. To go back to the rapist: it is human instincts he gives way to, not animal ones.

The popular image of a hyena is that of a mean, cringing creature who scavenges on scraps left over from their kills by bolder beasts such as lions, and on human corpses. Dr. Hans Kruuk, who lived among hyenas in Tanzania for several years, discovered that quite often it is the hyenas who make the kill, and the lions who are the scavengers. Hyenas, he found, are sociable animals living in communities, or clans, at peace with each other so long as each clan remains within its territory and does not try to poach, in which case its members are driven off, often with bloodshed; but, unlike human clans, or nations, defense really means defense and is not a synonym for aggression. The victor retreats into his territory. He does not put in an army of occupation, exact reparations or enslave the defeated hyenas. It happens, of course, as in all societies, that one group may become weakened by some untoward circumstance—disease perhaps, starvation, a natural disaster—to a point where it can defend its territory no longer and another, stronger community comes in. But animals do not wage war on other animals, as men do, for the sake of pride, aggrandizement or lust for power; only for the sake of safeguarding food supplies.

Perhaps the most striking divergence between the image and the reality concerns the wolf. In northern latitudes the wolf is feared and hated in a very deep layer of the mind. The big, bad wolf not only devours babies when thrown to him out of sleighs, but haunts the habitations of men, with long, sinister fangs ready to tear the throat out of any incautious human, and howls his deathly message to the moon. Fear of wolves goes back to the days when men dwelt in dark forests, and "werewolves" prowled abroad.

Some years ago, Farley Mowat was sent by the Canadian Wildlife Service to investigate allegations that wolves were ravaging the caribou population of northern Canada. He landed at a spot in the Keewatin Barrens with an arsenal of revolvers, rifles, tear-gas grenades and wolf-traps to protect him from the savagery of the ferocious beasts he was to risk his life to study. An Arctic summer spent in the constant company of a vulpine family, the parents George and Angeline and their four cubs, proved beyond doubt that the centuries-old image of the savage wolf was so much bunk. Nothing could have been more friendly, gentle, dignified and inoffensive than the conduct of the family to whom Mr. Mowat became so attached that he almost became a wolf himself, learning to take short wolf-naps instead of sustained sleep, and even sharing their diet which turned out to be, in summer, not caribou but mice.

The wolves' sexual activities, like those of the gorillas, were limited and chaste. The mating season lasts for only about three weeks every year, and wolves mate for life. Breeding can take place only when a hunting territory can be secured, and so a wolf may remain celibate for several years before he can find a mate and start a family. Wolves are much too responsible to plunge into parenthood, as humans do, before they can be reasonably sure of being able to support their young. As parents, they are impeccable, the male never failing to keep the larder full, save in times of famine, the female guarding, training and entertaining her young without stint. All the time Mr. Mowat was camped beside their den, seldom letting them out of his sight, never once did they portray the slightest sign of hostility. In winter, it is true, they did prey upon the caribou, but their role of predator, as in all animal communities, was to kill off the weak, diseased and imperfect and thus to strengthen the prey species by ensuring that only the strong and fit survive. It was men, not wolves, who were so ruthlessly hunting the deer that the caribou population of northern Canada dropped from an estimated 4,000,000 in 1930 to about 170,000 in 1963.

Called to investigate a report that hundreds of caribou slain by wolves were lying dead upon a frozen lake, Mr. Mowat did indeed find hundreds of dead caribou; slain not by wolves, however, but by a party of sportsmen landed there by plane to rake the herds with high-powered rifles, select from their dead and maimed victims the widest-spreading antlers, and return in triumph with their trophies of the chase. Yet the wolves were blamed, the men applauded, and the Canadian government paid a bounty for every dead wolf.

It would, of course, be foolish to pretend that no wild animal ever attacks a man. Almost invariably, in such cases, the man has attacked first and the animal's counter-attack is provoked by fear and pain. "Unprovoked" attacks, which are on record, generally have their origin in some former provocation. I recall one such case: a traveller's car stuck in a sand-drift, he got out to investigate and a rhino burst from the bush, charged and impaled him. The rhino was subsequently hunted down and killed. In his haunch was a deep, festering wound, crawling with maggots, inflicted by a tribesman's spear. A fair proportion of Africa's dwindling wildlife is suffering from wounds inflicted by the rifles, traps, spears and arrows of their human predators, and such "unprovoked" attacks are to be expected. Fear, as in humans, often stimulates aggression. Females defending their young are a special case, but even they are not invariably savage. Mr. Mowat, crawling down his wolves' den to take its measurements, and in the belief that its occupants were out, found himself face to face with two pairs of eyes gleaming faintly in the light of an expiring torch, his only weapon. It was Angeline with one of her cubs. He wriggled out backwards, sweating with fear, but he need not have worried. Angeline did not even growl.

Is it not time we changed our images, to bring them more in line with the truth? Timid, say, as a tiger, lazy as a lion, chaste as a gorilla, prudent as a wolf, sociable as a hyena, maternal as a crocodile, gentle as an elephant, faithful as a jackal, and inoffensive as a snake.

CIRCUSES AND ZOOS

My Wild Life

Jimmy Chipperfield

Most of us feel that, if animals are to be kept in captivity at all, it's infinitely kinder to keep them in conditions which come as close as possible to those of their natural state. At the same time there are few of us, whatever our age, who don't thrill to that heady music swelling out from the Big Top. It's a comfort to know that time after time really successful professionals of the circus insist that they would get nowhere with their animals if they didn't have a genuine affection and respect for them, though of course there are some regrettable exceptions. Through their first-hand experience with animals, too, circus people often come to know certain details about their "personalities" and behavior which would not normally be available to the more academic type of zoologist. One of the best-known names in the circus world of our day is that of Jimmy Chipperfield, and you only have to read his description of the training of lions and tigers to see how carefully he studied each one of them—and, indeed, how carefully he had to do so.

A circus lion's education generally begins when he is a cub. But he must not be too small: he must be big enough and have enough courage to challenge his handler. His natural instinct is to try the man out by rushing him and seeing if he runs away; if the man stands his ground, the lion is confused and does not press home his attack. I have seen some trainers so cool that they just sit in the cage reading a book and let the lions run about snarling and growling but not daring actually to attack. A very bad lion may come right at you at first, but if he does you push him off with a chair or some other solid barrier; that shows him you are the boss, and that you are not going to move out, no matter what he does. Once he understands that, the basis of training is established.

As with people, the distance to which you can approach varies from one lion to another. But whatever it is, once you go beyond the critical point the lion will strike out at you with his paw. If you press him closer still, there are only two courses open to him: either he must retreat, or, if he has not the room to do that, he must come for you. Half the art of training is to provoke the lion to a show of temperament that will excite the public, so that he roars and lashes out, but not to press him so that he feels really uncomfortable.

An expert trainer will go so near that he *does* make the lion come for him. He goes in really close, then as the lion starts to move he eases back, drawing the lion on, then stops again, and the lion stops with him. As he does this a few times he can see the lion thinking: "Right—in a minute I'm going to run him." And sure enough, in a moment the animal does try to run him, but after a couple of paces backwards he stands his ground and repels the animal again. In the ring the effect is terrific, but it can be achieved only after hundreds of hours of practice by a real expert. My nephew Richard is just such an expert: he can bring a lion bounding and snarling at him and at the last second turn his back on it—whereupon the animal drops its charge and lies down.

A critic might say that in this kind of training you are teasing the lion. In a way this is true. But to claim that it is cruel is entirely to misunderstand the lion's temperament. A lion is essentially an extrovert, and loves a rough-and-tumble. If he comes snarling at you, and you push him off with a chair, the odds are that he'll rush at you again, because he enjoys a bit of a battle. In making him swipe at you, you *are* teasing him, but in a way that he understands and enjoys. Far from being cruel, you're having a game with him, and he with you. A snarly lion is a good one: the more noise he makes, the better, and a good trainer will talk to him all the time, going *aargh* and *aaao* and making other lion noises. You can tell in a second from the sound a lion makes what sort of mood he is in, and even if your answers are not in very convincing lionese, the animal realizes that you are talking back to him and knows that you are a friend.

After a while the lion sees his trainer not as a human, but as another animal—and the boss animal at that. When the lion rushes the man, he is not thinking of killing and eating him, but merely of challenging him, to see if he can make him run. A lion who is *really* after you comes in a quite different way. He walks slowly. His head goes down, his jowl drops, and his face changes altogether. Far from running, he just keeps steadily walking. He is a different animal. An expert can spot the change in an instant, and many a trainer has been saved by his assistant, on standby outside the cage, noticing this change come over an animal and warning him in time. Yet, even if a lion does charge you only for fun, he is still dangerous enough, for he may well claw you and—if he draws blood—he

may lose his head and really bite, as did the one who killed Tom Purchase.

Another common misconception about lions is that they are all the same. Obviously their characters do not differ as subtly as those of humans, but all the same they show a considerable variation. Some are much cleverer than others, some more playful, and so on. They, in turn, see humans as individuals, just as much as we do, and it is always dangerous for someone to approach a lion—or for that matter, any wild animal—if he does not already know it. Once, after the performance of a winter circus, a man who worked with a tame tiger had the animal loose in his dressing room, and invited us in to see it. I refused to go anywhere near it, and the man was openly contemptuous, thinking I was frightened. I just told him that if he put the tiger in its cage, I would come and look at it, but not otherwise. How did he know the animal would like me? I asked him. What if it took exception to my clothes? What if my clothes didn't happen to smell right for it? How did the owner know that some small movement I might make would not set the tiger off? A really good trainer would never have issued such an invitation, but would have put the animal away first.

Another fact I soon learned about lions is that their perception is extraordinarily keen. Not only can they instantly spot a hole or a weak point in a cage: they can also sense a person's frame of mind from his physical movements alone. If their trainer walks into the cage firmly and slams the door in a determined way, they know they have met their match; but if a man comes in nervously, with a hesitant or faltering tread, they do not hesitate to try to mess him about. Even if the trainer only has a cold, or is a bit under the weather, it is risky for him to work, because the animals sense that he is somehow slightly different from normal, and therefore they become upset. This is the reason I have never drunk alcohol in any quantity, and never drink spirits at all: to go into the cage after a couple of drinks is the act of a madman. The trainer may think his voice and movements are perfectly normal, but the lions know better. Their sense of smell is acute, and they detect the difference at once—often with lethal results.

Tigers are even more highly-strung, and therefore even more dangerous. Except for the big Siberian tigers, which are relatively calm, they are difficult to handle, and you have to be tremendously careful not to hit them, even accidentally. If you so much as touch a tiger, he will usually lie down and refuse to work—and if he lies flat on his side, that is the worst possible sign. Anyone who did not know might think he had been knocked into submission and was lying there passively. In fact he is just about to come. His ears go flat down to his head, and the end of his tail flicks. As he lies there on his side, or even on his back, he is just a ball of muscle coiled and ready to spring: and when he comes, he uncoils and springs all in one movement, up to 18 feet in a single bound. So if ever a tiger goes on the floor, you don't go anywhere near him.

A good tiger, by contrast, makes a lovely sound—a kind of fluttering purr. You know at once if he's in a good mood, because he blows at you, and the trainer always talks back to him in the same language, blowing and letting his lips flutter. Some tigers never do this at all, being too nervous, but whenever one does do it, you can be certain he is all right. Even a good one, however, has to be treated quite differently from a lion, for no tiger enjoys the kind of rough-house on which a lion thrives. He needs altogether more delicate handling. If the lion is the carthorse of the big cats, the tiger is very much the thoroughbred.

From: *My Wild Life*, by Jimmy Chipperfield.

Jimmy Chipperfield also recounts the story of a publicity stunt, involving an elephant named Rosie, which nearly went wrong, but had a happy ending.

Rosie

The climax of Rosie's career came at the turn of 1930 and 1931, when she performed a star turn at the Piccadilly Hotel in London. By then we had a man who acted as our agent, and somehow the manager of the hotel heard that he had access to a tame elephant. The manager, it appeared, had conceived a novel idea for entertaining his guests in the evening of December 31 and for bringing the New Year in with a flourish: as the bells rang the Old Year out, all the lights would go out, and when they went up again, there would be an elephant standing on a tub in the middle of the ballroom, with a little Indian boy on its back. To follow the idea through, each of the guests would be given a model elephant as a souvenir.

Could we, the manager asked, provide the elephant? And we—whose policy it was never to refuse any reasonable offer or request—immediately said that we could. We knew that, Rosie's nerves being what they were, we faced formidable problems; but the man was offering us the princely fee of ten pounds, and for that nothing seemed too much trouble.

Had I known what we were in for, I would never have taken the job on. Nor would I do it now if someone offered me a thousand pounds. It was a miracle that we got through it without any serious mishap, and that we did not cause damage or an accident for which we could not possibly pay.

We were wintering in Shepherd's Bush, and we drove into central London easily enough, with Rosie in her trailer, and parked in some stables in a yard off Jermyn Street. All through that day we gave her very little to eat or drink, in case she should make a mess on the dance floor at the critical moment. As soon as she had settled, I went back to Shepherd's Bush, hoping to leave the whole business to Dicky. But suddenly I got an urgent message: the page-boy from the hotel who was going to play the little Indian rider had fallen off during a rehearsal and lost his nerve, so that I—still quite small at eighteen—was going to have to take his place as mahout.

I rushed back into town. The journey to the hotel passed without incident: we led the elephant along Piccadilly with her chains on, praying that she would not take fright at the traffic and bolt, and as soon as we reached the hotel I changed into Indian robes and a turban, making up my face and hands brown, and was given a huge basketful of beautifully made velvet elephants. Then we waited in a back room for our cue. At last midnight came. The lights went down, we trundled through into the ballroom and took up our position. A moment later the lights blazed up again, and we realized belatedly what we had taken on.

Everybody in that place was drunk. There was one great shriek of enthusiasm, and people surged forward at us from every side. I was convinced that Rosie would take off with the noise, but for once she was rock steady—perhaps because she was hungry and had seen and smelt all the food on the tables round the floor. Far from bolting, she just strolled forwards and began to filch things from the tables: rolls, fruit and sweetmeats disappeared by the plateful. Then she took hold of one of the tablecloths and with a twitch of her trunk sent everything on it crashing to the floor. Nobody seemed to mind losing the remains of their dinner: on the contrary, they were wild with excitement and began feeding her more and more delicacies. A few of them tried to climb on her back and she was knocking them down like ninepins with her trunk between mouthfuls. I sat tight on her shoulders, handing out the baby elephants as fast as I could. Soon someone began to shout: "Fetch her a drink! Fetch her a drink!" and while a tub was being brought she turned her attention to the potted palms, eating a few of them and throwing the others up in the air. Then a large tub appeared—a bath, almost—and into it went a stunning variety of drinks: champagne, wine, brandy, beer, whisky—anything that people could lay their hands on. The tub was filled to the brim, and Rosie (who was of course thirsty by then) drained the mixture off at one draught.

On her almost-empty stomach the brew produced an immediate effect, and although she kept on eating anything offered her, in a few moments she began to sway about. That was enough, Dicky and I decided: we had better get her home.

There cannot be many people who have had the task of leading a drunk elephant along Piccadilly in the first few minutes of a New Year; but those who have tried it will know what a terrifying experience it is. I don't suppose it took us more than a quarter of an hour to get her back to the stable, but the journey seemed to last half the night. The street was packed with cars and people, all of them merry, to say the least. They were happy enough already, and when they saw an elephant coming at them, that was all they needed: they all yelled and charged towards her. We hardly knew what to do: if we tried to go down the road, the drivers came at us blowing their horns; if we kept to the pavement, the drunks came at us on foot, blowing whistles and hurling streamers. And all the time Rosie just kept bowling forward, with Dicky and me each hanging on a chain, running with her, unable to stop her. Any moment, I was certain, we must run somebody down and trample them; any moment somebody was going to be killed. But somehow the elephant just kept clearing the bodies out of her way. Plenty of people were *knocked* down, but none of them seemed to resent it, and nobody was badly hurt. At last we were through the maelstrom and back in the calm of the stable, all three of us severely shaken. We got our ten pounds all right, but we agreed that it was the hardest-earned money we had ever come by.

From: *My Wild Life*, by Jimmy Chipperfield.

The Big Cats

Clyde Beatty

*Another famous trainer of lions and tigers (*trainer, it is always emphasized, never *tamer) is Clyde Beatty, who was at the pinnacle of his profession for more than 35 years. He has some exciting tales to tell of his experiences, and he makes it pretty clear that although he loved his big cats, he knew that it would never do to "trust" them in a human sense, for even the most seemingly gentle animal, he found, was capable of acting completely out of character—the character, that is, which humans had assigned to them:*

On January 13, 1932, I was giving my big cats a workout at the winter quarters of the Hagenbeck-Wallace Circus in Peru, Indiana. Nero, a big, powerful lion who had established himself as the arena boss, was about to go over a hurdle. Instead of making a clean jump he suddenly swerved in his course and came straight at me. It was one of those determined charges that an experienced trainer recognizes instantly. But though I knew it for what it was, I didn't have a chance to get set for it.

The first thing I knew I was flat on my back on the floor of the arena with the lion standing over me. It was the worst moment I had ever known. And I haven't experienced one quite as bad since.

As the big cat bent over me and bared the teeth with which he seemed about to mess up my features, I reached up with my right hand and planted it against his upper lip and nose. Then, with the strength born of desperation, I shoved him away from me, actually succeeding in working him back as far as my arm could reach. He gave his head a snap to release himself from my palm-hold, and as he did I found my hand in his mouth up to the wrist! This gave me a chance to gag him with my fingers. I needed that hand in my business and I was able to yank it out before he could recover his breath. The skin was scuffed where my hand and wrist had scraped against his teeth.

Nero did not make for my face again, but seized

what was nearest him. That happened to be the upper part of my leg. He grabbed it midway between the hip and the knee and tightened his jaws as if determined to snap the member in two. Having dug his teeth in deeply enough to satisfy himself (it developed later that they had sunk right into the bone), he began to drag me around the arena, bumping my head on the floor. Then he suddenly let me go, made for a nearby lioness and began licking her face with his tongue. The attack was over as fast as it had begun.

What saved me from being torn to pieces was the fact that he happened to move in the direction of a lioness that was on his mind. He had forgotten her momentarily in his determination to get me, but now that he was near her again he remembered that she was his main concern. The attendants outside the cage, frantically working their poles and yelling in an effort to distract the beast, had little effect on him. There was nothing else they could do. It would have amounted to suicide to enter the ring and try to fight him off.

It is my belief that the animal's attack on me was related to my having been standing near the female that had captured his interest. She was in heat and her presence excited him. In a situation of this kind an animal is capable of honest-to-goodness jealousy. If I had been another male lion, Nero could not have regarded my presence more suspiciously.

Later, Clyde Beatty tells of an incident while he was working with a tiger called Bobby on a Hollywood film:

Bobby was owned by animal trainer Louis Roth, and was known as "the wrestling tiger," one of Hollywood's most accomplished stunt animals. Bobby had been reared in captivity by Louis as a pet for his son, whom he had decided to launch as a Hollywood stunt man.

This looked like a good move for a man who was trying to find a career for his son, as there weren't many people in the animal branch of the stunt game. Roth, who knew the big cats as well as he knew bears, trained the young man to do close-contact work with Bobby suggestive of fierce struggles to get out of the clutches of the "demon," this "descendant from a long line of man-eaters." And the tiger, an intelligent animal, responded.

All big cats are dangerous, even those reared in captivity as pets by an expert like Louis Roth. But Louis had minimized the danger by making a one-man tiger out of Bobby—that is, it was unlikely that the animal would make any trouble for young Roth, to whom he had been accustomed.

During the filming, one situation developed that

necessitated the use of a wrestling tiger, and Bobby was offered to me, although originally the elder Roth had planned to make only "package deals" involving the services of both his son and the wrestling tiger.

I didn't like the idea of working with Bobby because he had never worked with anyone except Roth's son, and I made the point that the animal, like a one-man dog, would want no part of anyone except his master. But I was the star of this epic, and although the assignment didn't appeal to me, I agreed to undertake it. My experience with tigers told me that when Bobby found himself playing opposite me instead of his master, he would be hard to handle. And having just completed a pretty rugged circus season, I was not in the mood for needless problems.

The script for this serial involved a mythical jungle invented by a Hollywood screen writer who had devised his own particular version of the ancient formula in which you throw together wild animals, superstitious natives with weird tribal customs, a white man who gets into trouble and has to extricate himself, etc., etc.—and then hope for the best.

One of the big moments of the story called for my capturing a tiger in a camouflaged pit. It developed that in this particular jungle if a white man captured a tiger this meant bad luck for the natives. To placate the tiger the natives in turn captured me and tossed me into the pit with the animal—Bobby—as a sort of offering. If the animal had me for lunch this would pacify him and he would forgive the natives for their carelessness in letting me capture him. It was a very involved story and perhaps I'd better spare you the rest of it.

To foil the natives and save myself from the "murderous beast" that was bent on "devouring" me, I was supposed to wrestle him into submission, then put him in a trance with the magic of my hypnotic eye.

I hadn't wrestled with Bobby very long when it became apparent that my original hunch was correct and that I was going to have trouble with him. Louis Roth, perceptive animal trainer that he was, had decided to stand by for possible emergencies; he could see that things were not going well and called to the animal from above in order to get the obviously tense and nervous creature to relax.

Bobby, in unfamiliar surroundings and with a new wrestling mate for the first time, became over-excited and to show how he felt about things bit me in the shoulder. He could have removed my whole shoulder if he had been in a better position to get set and give his jaws full play. In fact, he might have killed me. Anything can happen when a so-called "tame" animal panics. He had me against a wall of the pit and was struggling to manoeuvre himself into a position for a fuller and more authoritative bite when I hauled off and clouted him behind the ear with everything I had.

Bobby, who was accustomed to being petted, not socked, was shaken by my wallop. It was a new experience for him. Before he could decide what to do about it, the excited animal looked up and caught Louis Roth's eye. Louis saw the bloodstain spreading over my shirt and beckoned the tiger out of there. Bobby complied. Roth had trained him from the beginning, and Bobby recognized him as the final word, just as he accepted Roth's son as his sparring partner.

Breezy Eason, the director, acting as if nothing had happened, cupped his hands and yelled down to me after Roth had got Bobby out of the pit, "In a few minutes we'll do it again. That was pretty good—almost the way I want it."

From: *Facing the Big Cats, My World of Lions and Tigers* by Clyde Beatty.

A Circus Year

Michael Marden

Amateurs sometimes enter the circus world also. Michael Marden, for example, was an ex-public schoolboy and Cambridge University student, who for a time became Cuthbert the Clown. He also made friends with Carlos Rosario, who had a pigeon act—and a love for animals of all kinds.

Carlos, who was a Mexican, was a "character." He had been the champion rope-spinner of Mexico in his youth, and had also, at one time, been a bull-fighter. He had an extremely fierce cock bantam called Charlie —or Yarlie as Carlos called it—which used to fly at everyone who entered the caravan and peck their leg savagely. It slept with Carlos in his bed, and refused to allow anyone near him. Carlos thought the world of it. It crowed incessantly.

In addition to Yarlie, Carlos had a tame jackdaw. He had picked it up in the road—it had fallen out of its nest—and it had grown into a bird of the greatest sagacity and charm. It delighted to play catch with Carlos, using little balls specially made for it by Mrs. Carlos out of chamois leather. These it would catch in its beak and throw back to Carlos quite happily for hours. It could also do simple card tricks. One of these consisted of Carlos laying out four cards on a ledge of wood originally designed for Mah Jongg, and saying, for example, "Bring me the nine of diamonds, Blackie." Blackie would then hop along the table, examine the cards, pick out the right one, and hop back with it in its beak to Carlos. The jackdaw was extremely possessive, and couldn't stand the dogs, the pigeons or Yarlie, and spent a great deal of its time abusing them. It was devoted to Mrs. Carlos, and used to perch on her shoulder and smother her with kisses whenever she returned after having been away five minutes. Its demonstration of affection over, it would then steal her

earrings and depart. It would spend its days strutting about the roof of the caravan, regarding the scene with a gleaming and ironic eye, and making caustic comments on the world in general and Yarlie in particular.

Blackie was, in fact, in love with Mrs. Carlos. He tried to make her join him in his selected nesting site above the bookcase, and was assiduous in feeding her with pieces of cake, or, when the occasion offered, with a beak-full of minced worms well mixed with saliva, which he would force attentively between her lips. Poor Mrs. Carlos did her best to avoid these tid-bits, but Blackie was nothing if not persevering in his attentions, and not for anything would Mrs. Carlos hurt his feelings by spitting out his offerings!

Carlos seemed to attract unusual animals and once told how a hunted fox had leaped through the open door of his wagon and on to his bed:

The hunt had come round, and Carlos had told them he had seen the fox making off in a certain direction; the field had ridden off in pursuit. The animal stayed quite happily with Carlos and had apparently been intrigued when he played the mouth-organ to it. That evening he had taken the fox—or vixen, as it turned out to be—to the top of a hill, near a little wood, and had set it free. The fox is not, in the nature of things, a docile animal, but this one had lain quietly in Carlos's arms until he put it down, when it trotted off a few paces and then, as Carlos said, "She kept stopping and looking back, just as much as if to say 'Well, goodbye, I'll be seeing you later maybe.'"

From: *A Circus Year*, by Michael Marden.

The Old Showmen Knew a Few Tricks

George Sanger

One of the greatest of all circus men was George Sanger —or, rather, "Lord" George Sanger, as he was always known. He resigned from the business in 1905, and ten years later published Seventy Years a Showman, *one of the most fascinating books about the profession ever written. He, too, was a cool and resourceful lion trainer. He was at the peak of his fame in the 1870's, both in Britain and on the continent, where he made many highly successful tours. In 1876 he was with his show in Paris. The proprietor of one of the theatres there was about to produce a stage version—or rather free adaptation—of Jules Verne's famous book,* Around the World in Eighty Days. *He had seen one of "Lord" George Sanger's acts, which featured an African lion trainer and eight lions, and conceived the bizarre idea of introducing them into his play in a love scene which was supposed to take place in an African forest. He engaged "Lord" George to arrange the scene for him. As he didn't want to deplete his travelling circus of lions,*

Sanger sent to England for another eight which he had there in reserve. They arrived at the theatre in special wooden cages, which were lowered into a cellar under the stage, in readiness for the performance of the play on the Sunday.

Lion Roundup

I was very early astir on the Saturday in order to rehearse the lions before very many people were about. Directly I got in sight of the theatre, however, I was astonished to see a crowd about it. As I drew nearer I could see there were a lot of gendarmes present, and also my two men from Margate. When they caught sight of me they rushed forward with faces white as wax, Stratford wringing his hands and crying: "Oh, Guv'nor! Guv'nor! The lions are loose!"

"Loose!" I exclaimed; "what do you mean?" "They are loose from their dens," he replied, "and this gentleman here," pointing to a gendarme officer, "says they must be shot in the interest of public safety."

"Oh no," I said to the gendarme, "no shooting, please." Then, turning to my fellows, I said: "Come along! Come along! Let us get them into the dens!" To my surprise they did not budge. "Come along!" I said again; "aren't you coming?" But I got no response, so with a few unkindly remarks as to their want of pluck, I took the oil lamp from the watchman, who had been on duty at the theatre, and told him to unlock the stage-door.

When he had done so, I entered alone, the oil lamp in one hand and an ordinary walking-stick in the other. I rambled all over the theatre, stage, dress-circle, pit, etc., and finding no trace of the lions concluded they were still in the cellar. With the dim light I had it was difficult to find my way about, but down I went, and not seeing them in the upper cellar, crossed over to descend to the lower one. As I did so a lion suddenly made a rush for the same opening, and as he came, struck me with his head in the small of the back with such force as to make me turn a complete somersault.

I landed on my feet, thanks to my old circus experience, but I confess that for the moment I was unnerved. The lantern, however, was still in my hand, and still burning, so after collecting my thoughts I descended the steps to the lower cellar. Then I made for the spot on which the dens had been placed. There was a great deal of old scenery, rubbish, and cast-off properties about, so I very carefully made the round of the cellar, picking my way at every step.

All at once I saw eyes like balls of fire in the distant darkness. "Oh, there you are, you rascals!" I shouted, knowing that the animals would recognize my voice. Then I struck my cane on the various properties lying about, and at the same time swung the lantern to and fro. This had the effect of making the eight lions leap and bound in all directions. The rattle of the old canvas and other material that was thrown over by the heavy beasts, together with their surprise at my appearance, made them run round the cellar several times. By this time I was quite awake to the situation. I knew from experience that the beasts would make for their dens when they tired a bit. So it proved, for presently after another race round they made for the cases they had escaped from. I saw three get into one of the great boxes, and five into another, leaving two empty. Then I pushed to and blocked as well as I could the sliding doors of the cases, and hurried up to inform my men that the danger was over, and the lions were safely housed.

"Lord" George Sanger wasn't called "the great showman" for nothing. He knew and practiced all the tricks of the trade. In his earlier days, when he was travelling round Britain from one fairground to another, a very popular act was one in which a pig apparently displayed the most remarkable intelligence and wisdom:

The Learned Pig

Well, now, the making of a learned pig is upon this wise. You get your pig, fat and comfortable-looking and not too old, a fairly long stout stick, a leather strap that will buckle neatly round the pig's neck and has also a small plate and screw rivet that will attach it to the stick. Then you are ready to commence the lessons.

In the end of the stick, not the end to which the strap is attached, you bore a hole, and through this drive a long nail into the floor of your academy so that the stick can move freely round on it in a circle, but in no other way. When the pig's neck is buckled into the strap at the other end of the stick the animal is bound to move in a circle, of which the nailed end of the stick is the centre.

Then with a little cane to direct his movements you induce the pig to walk. Of course he goes round and round and round, for he can move in no other direction, and when he wants to stop, which is often, you just keep him going by gentle taps with the cane. When you have kept him walking round some time you begin to let him stop in his course now and again, but always just before the stop giving a slight click with the fingers. The slightest sound will do, merely the snap of the thumb-nail against the finger-nail is sufficient. The pig will hear it, and in a very short time will stop this monotonous walk as soon as he hears the slight signal.

You then vary the lesson by arranging a pack of cards face upwards just outside the circle, fixed, of course, by the length of the stick, which the pig traverses, and commence to patter as if to an audience somewhat in this style: "Well, Toby, you see the cards before you. Which is the ace of spades?" Any card you like you can, of course, name. Round goes the pig in his circle, and as he comes opposite the card "click" go your nails, and he at once stops.

"You see, ladies and gentlemen," you proceed, "Toby knows the cards. Will someone kindly name a card they would like him to pick out." Round goes piggy as you patter, and "click," you stop him where you like. In two or three days the pig, without the stick or the strap, will commence to move round at a tap from your switch whenever a circle of cards or persons is formed. He will also stop dead at the finger-click until the touch of the switch lets him know he must move on again. Then his education is complete.

You can send him round a circle of people, asking him to pick out the man that likes kissing the girls. In fact, vary your entertainment as you will, the pig will be listening for the "click," not to your patter, and will stop as soon as he hears it, while the audience will not notice the slight sound. With every performance the pig will improve, especially if you accustom him to receive after each show an apple, potato, or some such little luxury.

From: *Seventy Years a Showman*, by "Lord" George Sanger.

Zoo Babies

Paul Steinemann

In many cases it is only at a zoo that the birth and infancy of certain species of animals can be closely studied. Here, for example, are two stories from the Zoological Gardens at Basle, in Switzerland. The contrast in the development of the babies at birth is extreme.

A Kangaroo Family

"It looks very much as though we are going to acquire an addition to the kangaroo family," Keeper Glücki announced in November 1953, after he had noticed movements in Adelaide's pouch. We were overjoyed, for our kangaroos had remained childless over a number of years, but we were to have a double surprise for Darling too harbored something suspicious in her pouch, and before long we ascertained that both females had had babies. It is an incredible feat that a newborn kangaroo reaches its mother's pouch, for at birth it is only half an inch long and weighs just a few ounces. It really is an embryo still, and will fit comfortably in a teaspoon. Keeper Glücki had seen a female pick up her tiny child gently with

her lips, open her pouch with her two front paws, and put the infant inside, but what it does there no one knows. Probably it just feeds and rests until it is strong enough to venture outside, but during the first week of its life it is not strong enough even to suck from its mother, so she has to release her milk herself and direct it into the mouth of her child. Neither Adelaide nor Darling allowed us to peep into their pouches to survey the development of their children, and we could only guess from the increasing bulk that inside the pouches the babies grew and moved around. Only their father was permitted to put his head into the cradle to make sure that his offspring were developing properly.

After five months our patience was richly rewarded though when a prim little nose and two enormous eyes peeped out of Adelaide's pouch and then promptly vanished. But it was at least another month before the baby dared to climb out of the pouch and tried its legs for the first time. Darling's child waited even longer and did not venture out until the end of May. The demure little beast, hardly bigger than a rabbit, stepped out composedly and then rushed across the enclosure and jumped into Adelaide's pouch. Their keeper was most surprised to see two heads emerging from the same pouch. Adelaide was not at all disturbed by this sudden increase in family but Darling was most upset and whimpered while looking everywhere for her lost baby. Confused, she hopped round the cage until eventually, thinking that the smallest of

BROWN BEARS

G

RING-TAILED LEMUR

the other kangaroos belonged to her, she tried to put it into her empty pouch. We had to rush to its rescue, and then carefully lifted both babies out of the pouch of the resisting Adelaide, compared them with their mothers, and gave to each the child which resembled it most. We did not have any opposition to this move, and soon peace and quiet were restored to the kangaroo enclosure.

The infants spent longer periods outside the pouches, and became more daring in their excursions. We noticed that each baby used the absence of the other by hurrying to its aunt and gobbling her milk from the pouch. (I myself took the opportunity of any empty pouch to investigate its contents, and put my hand inside where I found four distinct nipples.) The mothers did not seem to mind this robbery—on the contrary, they encouraged it by licking the feeding child, regardless of its parentage. By September the children were too big to get into the pouches, but they fed from their mothers even after they were a year old.

Father kangaroos take little interest in their children, but they guard their wives jealously and challenge anyone who approaches them. As soon as I entered the cage the father would hop towards me, stand up and playfully make boxing gestures. If I accepted the challenge he would straighten himself and take up a fighting position. If this did not impress me, he would get really angry and try to kick me in the stomach with his powerful back legs. To measure the force of such a blow, imagine a kangaroo springing ten feet into the air or leaping twenty-seven feet along the ground. In addition to its leg muscles, the kangaroo has sharp claws on its feet, which assist it in running, and these too can be dangerous. Fortunately kangaroos are generally harmless creatures and are loving and trustworthy to their keepers.

Birth of a Giraffe

In 1953 we expected to have a baby giraffe for Christmas since Susi was already fourteen months pregnant and approaching the day of her confinement. She was under constant supervision but Christmas came and went, and the New Year superseded the old, and still there was no sign of the uncommon infant. Then, on the afternoon of January 4, the alarm was given, and we hurried to the antelope house, where the birth was already in progress. Giraffes give birth standing up,

and after a tiring half-hour the newborn calf somersaulted to the ground. He lay quite still for a few minutes, then raised his long neck, listened in all directions with his neat little ears and made his first attempt to stand up. His long legs were not yet strong enough though, and he fell back on the straw, but he was not deterred and before he was half an hour old he succeeded in balancing gracefully on his long thin legs. Immediately be began to look for food, and when he eventually found Susi's udder he sucked for thirty minutes. He was a healthy, well-developed male calf, Susi's second child, weighed about 110 pounds, and was five feet tall. When he was only two hours old he pranced round the stall, and when he lay down his mother licked him lovingly and took care not to tread on him.

We were relieved that Susi treated her baby so tenderly, for with her first calf, her mothering instinct had not developed, and she had stood ignorant and helpless beside it. Unfortunately, the baby was premature and was only four feet high and could not reach its mother's udder. Susi hardly licked it at all and seemed to be afraid of it. She looked at it suspiciously, and then carelessly trod on its leg, which broke. We had to take the calf from her and bottle feed it. We gave it over a gallon of cow's milk a day, sweetened with banana pulp and containing vitamins and calcium; even the rubber teat was coated with the sweetened preparation, for giraffe children are extremely fond of sweetmeats. Sabinchen, as he was called, became the favourite of the visitors and he was the friendliest of beasts. He adored the keeper who had replaced his mother, and licked his face affectionately. He thrived in our care but his weakened leg did not mend as quickly as it should have. He was just beginning to make progress when he injured another leg severely and so, in spite of our attachment to him, we had to end his misery.

Happily Susi had better luck with Baschi, as we called her second child. She looked after him attentively, and if she thought his safety was threatened by visitors, she rushed to his defense. She placed herself between him and the public, and for the first few days not even her trusted keeper was allowed near him. Indeed, this fear seemed to supplant even her love for him, for when we let Baschi go into a larger stall to which he was drawn by curiosity, Susi was so nervous that she could not be persuaded to cross the threshold even at the risk of losing her baby. Eventually we had to take her into the stall by another entrance to join Baschi.

In the spring we allowed Baschi to go out into the paddock. He galloped around with delight and made the acquaintance of his father and aunts. Father and son scarcely took any notice of each other, and it was soon apparent that the child could scarcely distinguish between the grown-ups, for he mistook each of his aunts for his mother and tried to feed from all of them in turn. He only realized his mistake when they were uncooperative.

Baschi grew quickly during his early months and when he was two months old had grown nine inches; during the next four months he grew another seven inches and at the end of his first year was over nine feet tall. In their natural state giraffes are about eight and a half feet tall measured at the shoulder, which Baschi reached when he was ten months old.

The most impressive feature of a giraffe is, of course, its long neck which is made up of as many vertebrae as the necks of seven men. Chewing the cud interests visitors too, for the stomach ejects a ball of food which is visible as it travels up the neck to the mouth. It is chewed about forty times and then swallowed again. The favourite food of the giraffe in the zoo, as on the African Steppes, is the twigs and leaves of the acacia tree. I am constantly amazed by the way in which the beasts swallowed this thorny fodder without the least discomfort, but their long black tongues which curl round the thorns seem to be immune to their pricks.

Fortunately giraffes are not malicious creatures, but they could, if they wished, throw a man to the ground with one swipe of a foot. Their enormous eyes give them a peaceful air, but one must beware of frightening or exciting them for they are prone to panic, and with their strong legs can easily defend themselves, even against lions. Their tiny skin-covered horns are also used as weapons, and by shaking their necks vigorously from side to side they can effectively hurt their opponents. This method of defence is occasionally used against their keeper, if they consider him a potential rival and want to force him out of the enclosure. They can be very stubborn and have to be shown who is the master. Achmed, our bull, knows his keeper well and knows exactly how far he can go with him, but in spite of his appearance, he is a nervous beast and can be terrified by an unfamiliar occurrence, no matter how trivial. Once his keeper was replaced for a few days by a substitute who wore spectacles. Achmed dared not approach him, not even when offered a bucket of maize, until the reason for this strange behaviour dawned upon the keeper, and he removed his glasses. Then the animal walked up to him and ate quite happily. Obviously a keeper has to have infinite patience and affection for his charges to gain their trust.

From: *Cubs, Calves and Kangaroos*, by Paul Steinemann.

Victor

The most famous giraffe of recent times was Victor, from the Marwell Zoological Park, near Winchester in the south of England. Unfortunately his story did not have a happy ending.

When Victor's keepers arrived for work one morning they found the fifteen-year-old East African bull giraffe spreadeagled on the ground in his night enclosure, his front legs folded and his hind legs splayed out behind him. With varying degrees of coyness, the papers suggested that Victor had been making "amorous advances" to one of his three female companions and had had the misfortune to topple over, but the zoo keepers say this is pure speculation. The fact is that a giraffe that does fall in this way is in a very awkward predicament, because it's extremely difficult for the spindly-legged creature to lurch to its feet again. In fact, if it happens in the wild the giraffe may simply lie there until it dies or, more likely, falls victim to a predator.

Victor was a full-grown bull, a prize stud and the father of seven calves. He measured all of eighteen and a half feet tall to the top of his head and weighed a ton. So it was hardly a simple matter to get him on his feet again.

The fruitless efforts of the zoo staff and their helpers to lift Victor to his feet occupied the newspapers for days on end. It was soon established that the giraffe, who seemed quite unconcerned at his predicament, was physically undamaged. Eventually, after several hours of unsuccessful endeavor, the zoo men gave up and did what most people do when confronted with some insuperable physical problem: they called the Fire Brigade.

A dozen stalwart firemen, complete with fire helmets and brass-buttoned tunics, approached the task with enthusiasm. What about a crane, someone suggested. It took three hours to put the one-ton stranded giraffe into a huge canvas harness, with its two giant pairs of trouser-like slings that had been specially made to fit him. At last, the time came to start the lifting operation. The four winching handles of the lifting gear were slowly turned to raise the animal's enormous bulk, suspended between four tall scaffolding gantries. A crowd watched quietly, mindful of the paramount need not to upset the temperamental giraffe. Victor, struggling slightly, was comforted by his special keeper, 21-year-old Ruth Giles, who was perched on bales of straw beside him. As the giraffe gradually rose from his collapsed position, zoo staff massaged his legs to try and get his circulation going. The hope was that the hoisting operation would enable him to be supported in a standing position while his legs regained sufficient strength to bear his weight. But it was not to be. The strain seemed to have been too great, and within minutes of being hoisted upright Victor died, in a vet's opinion, from shock.

So ended a six-day saga that aroused sympathetic concern all over the world.

From Circus to Game Park

Jimmy Chipperfield

Jimmy Chipperfield was a pioneer and enthusiastic supporter of the whole concept of the game park in which animals could be displayed to the public in a spacious setting that approximated their natural habitats. In 1964 he met the Marquess of Bath and discussed with him the possibility of opening such a park on his estate at Longleat, in Wiltshire. Later they became business associates in the running of the enterprise. The experience taught him many lessons about animals.

Our animals have put to rest many of the myths which used to shroud the business of reproduction in mystery. Many zoos, for instance, failed to breed chimpanzees, and the idea got about that chimps had to be shown what to do before they would mate properly. No greater piece of nonsense has ever been believed, as was demonstrated by Charles, a tame chimpanzee whom we had reared in our house and put on to the island in the lake at Longleat.

He first distinguished himself by rowing off, on his own, in the boat belonging to two of the park staff who had come to visit his island, leaving them stranded. But his real triumph was that he immedi-

ately mated with the wife we had given him—a chimp who had lived in a zoo and was altogether different from him in temperament, having never been close to humans. According to traditional theories, they were as unlikely to produce a baby as any pair we could have devised, yet they managed it. The only unhappy consequence was that when Charles's wife did give birth, she abandoned the baby, probably because she had no milk. We rescued the infant, christened him Fred, and brought him up on the bottle.

We also brought up a chimp which had been born in one of the parks. Ours, which we called Simone, was the daughter of Alby and Mona, two very old animals whom we installed on the island at Blair Drummond. Mona has astonishingly strong maternal instincts, for although she never had any milk she carried her infants about clutched to her chest for months after they had died. For six months, in one case, she clutched the pathetic little black, dehydrated stick to her—a heartrending sight. Five babies were born and died like this, and we kept hoping that a miracle would occur and her milk would come; but when the sixth arrived we could not bear to watch the tragedy happen again, so we anaesthetized the mother with a dart, took the baby away, and reared it at home. Oddly enough (for both her parents were hideous) Simone turned out very pretty, and altogether is a charming animal.

Another myth which we scotched was that hippos would never breed in a cold climate. Our first hippo, Arnold, took to the lake at Longleat as though he had known it all his life, and thought nothing of breaking the ice in the winter; nor did the cold in any way hamper his sex life—he began mating when we thought he was still much too young. Unfortunately, although a highly attentive husband, he proved a disastrous father, for no sooner was his first baby born than he attacked and killed it. Since then he and his mate have produced two more offspring, and in both cases we have had to remove the baby and rear it by hand to prevent a repetition of the infanticide. As far as we can tell, it is jealousy of the third party that makes Arnold so vicious, but we have found no way yet of curing it.

Altogether he and his mate gave us a considerable amount of trouble. At first they were fenced into a small enclosure on the bank of the lake, but Lord Bath became keener and keener to let them out, and I decided it would be all right to do so, thinking that they would not wander far, even though the lake as a whole was not fenced. I was proved entirely wrong, for they began wandering miles in the night, just as hippos do in Africa, and Roger, at the park office, kept getting agonized telephone calls at all hours. Sometimes it was the gardeners from Longleat House, complaining that the hippos had been through the rose-beds again, and sometimes it was customers at the Bath Arms, a mile up the road, saying that even though they admitted that they had had one or two, they still thought we ought to know that a hippo had just sauntered past the window. The animals—and their tracks—were so enormous that when damage was done there was never any point in trying to pretend that some other creature had been guilty. Eventually we decided that there was no alternative but to fence the entire lake.

Over the years the parks have given us many fascinating insights into animal behavior. The lions, for instance, soon realized that although the grassland of the park was their territory, the roads belonged to the cars. As long as the cars stay on the roads, no lion takes the slightest notice of them—as anyone who has driven through one of the reserves will testify. But let a strange car go a few yards off the paved road, and immediately some of the lions attack it, chasing it, biting at the tyres and striking out at the mudguards. I had noticed this already in the south, before we

opened at Blair Drummond, but the truth of the observation was brought home to me when Princess Anne visited the Scottish park. I drove her round in a brand-new Land-Rover, and to give her a close-up view of a lion I went on to the grass, towards where a group was lying. At once they leapt up, regarding us as an intruder, and one charged up against the front of the vehicle with a thump. The Princess was never in any danger, but the incident certainly livened up her morning.

And of the animals' attitudes towards their keepers, he writes:

A striking difference has become apparent between the attitude of the carnivores towards their keepers and that of the hay-eaters, like giraffes and antelopes. Whereas the lions and tigers regard their keeper with respect, and accept him as the boss—just as they would in a circus—the hay-eaters are definitely jealous of the people who look after them. The one fatal accident which has happened in any of the parks occurred at Woburn when a bull eland gored his keeper as the man was trying to herd him in for the night. The eland had no intention of leaving his females, and when the keeper tried to move him on by flapping his jacket, the animal darted at him and with one thrust of his horns flicked him up in the air. Although the man managed to get back to his vehicle, he died inside it, gored all through the middle. That animal had never given trouble before; but, like any bull, he was touchy about his females, and his attitude (I am sure) was very much like that of a farm bull, who can often be more of a menace to the farmer who looks after him than to anyone else. Like the bull, an eland or giraffe develops a particular jealousy of his keeper, and is irritated by the fact that the man is always loitering about, moving him on and challenging him (as he sees it) for the supremacy of the herd.

I am certain that after a while he begins to say to himself: "Why's this person always hanging around? Why doesn't he (or she) go away?" The result is that whereas in the lion and tiger sections the main threat would be to the members of the public if they got out of their vehicles, in the antelope and giraffe sections the public is perfectly safe to mingle with the animals, and it is the keepers who are at risk.

The behavior of the giraffes at Longleat has demonstrated this perfectly. Several of the females have given birth in the middle of hundreds of visitors, taking not the slightest notice of all the spectators. But one bull giraffe, who ignores the public entirely, has several times chased his keepers with vicious intent. Once he went for the girl looking after him and chased her round and round a tree, trying to knock her down, until she was forced to call for help on the radio. The moment she was picked up and taken off in a vehicle, the animal calmed down and returned to his grazing among all the picnickers. Again, I have seen visitors in several of the parks taking pictures of each other with their arms round the neck of one of the Ankole bulls—huge creatures with enormous horns nearly five feet across. The animals do not seem to mind in the least. Yet if their keepers tried to take any such liberty, they would knock them down in a flash.

From: *My Wild Life*, by Jimmy Chipperfield.

Temba and Tombi: A New Breed of Lion?

Chris McBride

In the Lowveld of the Transvaal in South Africa, not very far from the western boundary of the huge Kruger National Park, and some 350 miles northeast of Johannesburg, there's a private game reserve called Timbavati, after one of the rivers that flows through the region—when it *does* flow, that is, for during much of the year it's not much more than a wide expanse of soft sand. The reserve came into being in 1955, when twenty-eight local landowners agreed to combine their holdings in the Lowveld in order to create the conditions for the preservation and fostering of their wildlife, in the spirit of King George VI's declaration:

"The wildlife of today is not ours to dispose of as we please. We have it in trust. We must account for it to those who come after."

The Timbavati Nature Reserve consists of 208 square miles of virgin brush. It's part of a very ancient area which has experienced hardly any geological changes—no volcanic eruptions, no ice age, and no glaciers. And there have been very few man-made changes either: there are no fences, for example, and the parcels of land within it are marked off only by tracks cut through the bush, while the original holdings were never cultivated but used by their owners only as hunting-grounds or country retreats, and are now almost entirely devoted to the conservation of wildlife. So Timbavati is in many ways ideally suited to its present purpose, and it teems with wildlife of all kinds . . . including lions.

Early in 1975 a young man named Chris McBride moved into his father's old holding in Timbavati, with his wife Charlotte and their small daughter Tabitha, in order to fulfil the fieldwork requirements for a master's degree in wildlife management at Humboldt State University in Arcata, California. For his thesis he decided to study in depth one particular pride of lions. He called it "the Machaton pride," after another river which ran through the pride's territory—though this river, too, was dried up for most of the year.

The Machaton pride consisted of (and perhaps still does) two superb males, named Agamemnon and Achilles, who probably came from the same litter and may even have been twins. At any rate, they get on remarkably well together, sharing between them the six lionesses of the pride in a perfectly amicable fashion. The names Chris McBride and his family gave to the lionesses were: Golden (because of her color), Dimples (because she had a black mark like a dimple on one cheek), Scarleg (one of her legs was scarred), Greta (after Greta Garbo, the filmstar who was always saying that she "wanted to be alone"), Lona (who was another loner), and the one who matters most in this true story, Tabby, named after Chris McBride's daughter Tabitha. In addition there were Golden's three young sons, known as the Three Musketeers, and their sister, Suzie Wong.

For several months the McBride family and their helpers tracked the Machaton pride in their jeep, patiently studying it, until the lions came to accept the presence of the vehicle without suspicion—probably, Chris McBride thinks, because the smell of gasoline, rubber tires and so on masked the scent of humans. Then on June 6, 1975, Chris McBride and an American friend, Joe Zamboni, who was staying with him, were accidentally present (though, of course, at a safe distance) at the mating of Agamemnon and Tabby. This in itself was quite a rare event for humans to witness. But the result of the mating was to be far more amazing:

⟡

We found the cubs completely by chance in early October 1975. Well, not completely by chance. Charlotte had seen Tabby in mid-September looking extremely pregnant. Two weeks later she again spotted the lioness, looking much more slender. It seemed likely that Tabby had had her cubs, and Charlotte told us roughly where to look for her.

The weather wasn't very good that day, and as I was suffering from flu, I decided to stay in bed and read. My older sister Lan, up from Johannesburg for a visit, went off in the Wagoneer with her son James and Johnson, one of the older trackers in the area. Johnson, a Matabele, is a very rugged character. He's been known to walk all the way from his home in Plumtree, Rhodesia, to Johannesburg, a good 450 miles, most of it through lion country. He'd walk all day and sleep

up in the trees at night; the journey would take him three or four weeks, but he thought nothing of it.

Hoping that Tabby had already given birth and that they might catch her in a rare moment with her new cubs, Lan, James and Johnson headed straight for the place we called the Plains, an area of sparse scrub and knobthorn trees.

As they approached the spot that Charlotte described, she suddenly braked and switched off the motor.

"Look, James, a lioness," she whispered.

"Where?" he asked, searching the terrain without any success. Lions are notoriously difficult to spot, even when you are right on top of them.

Then he saw her. It was Tabby, only about 25 yards from the track, under a shingayi. She lay there, quite unconcerned, looking back at the jeep. Beside her was the remains of a wildebeest kill on which she'd been feeding.

In absolute silence, Lan, James and Johnson sat and waited.

Within a minute, a little head popped up behind the lioness. To Lan's amazement, it was snow-white.

Then another little head appeared—tawny.

And yet a third—snow-white again.

Lan was staggered. She'd heard of very rare albino cubs, but these had yellow eyes—normal lions' eyes. It was quite clear that they weren't albinos. They were ordinary, seemingly healthy lions. Except that they were pure white.

She drove back to get me—"like a maniac," as she put it. She found me still in bed, reading and unable, because of the flu, to get as excited as I should have been about the news.

"Are you sure they're not just pale?" I asked her.

She was furious. "No. They're white. Pure white," she insisted.

By this time I was sufficiently interested to drag myself out of bed and grab a camera. I was only sorry that Charlotte happened to be helping out in the Sohebele game-viewing lodge, too far away to be fetched in time, because at this period she knew far more about the workings of this particular camera than I did.

So, in spite of my health and my own basic doubts about any such thing as white lions, I joined the crowd around the Wagoneer—which by this time had become quite considerable. The news about the white lions had quickly spread around the camp and a rather impressive viewing party had assembled.

Lan drove back to the spot where she'd seen Tabby and the cubs. They were all there, still in the same position. The cubs stayed close to their mother, but now and then we would get a glimpse of one of them over her tawny shoulder. But only a glimpse. Never once did we see all three cubs together. I worked away with the camera, as best I could.

After twenty minutes the cubs grew bolder. One by one they wandered off into the bush behind their mother. We now spotted another lioness, Dimples, lying in the nearby shade. Tabby stood up, and from the Wagoneer we could hear her calling softly to her brood. It's hard enough to indicate human accents phonetically, let alone animal cries, but it was a short, sharp sound, something like the *oo* in *book*, but higher in pitch. This was the seldom-heard sound that adult lions make when they are hunting together and want to communicate without alerting other animals. It's a sound that has evolved over the centuries, a sound which is nondirectional and doesn't seem to attract the attention of other animals.

Tabby called again—the same short, sharp *oo* sound —and yet again, a third time.

From a thicket emerged a tiny tawny form. Then two more furry shapes—white as polar bears, but unmistakably lions. It was the first time I had seen them in the open. One had broad, quizzical features, suggesting an adventurous male. This was Temba, as we came to call him: Temba, the Zulu word for "hope." The other had the triangular face of a female and was much more cautious. She became simply Tombi, which is Zulu for "girl."

Temba and Tombi ambled over to join their brother, whom we later named Vela, meaning "surprise." And there they all stood—Vela and the white lions of Timbavati—in the secure patch of Lowveld grass, listening as their mother called out to them.

Thirty yards away, Lan started the engine and we turned discreetly for home, leaving them to the wildebeest carcass that had so fortuitously pinned them to the spot.

Looking back on it later that night, I began to realize what a momentous find we had made. The reality of having discovered—no less photographed—the first truly white lions in recorded history was quite overwhelming to me.

⁂

There was no doubt about it, Temba and Tombi really *were* "as white as polar bears." There have been quite a number of cases of albino lions—there's one in the Kruger National Park in fact—and, if it comes to that, of other animals such as tigers and leopards. But albinos are the result of an abnormal and total lack of pigmentation, usually affecting the eyes—which are often pinkish in color—as well as the rest of the body. The white lions of Timbavati are *not* albinos: they are positively white-pigmented, and their eyes are the yellow of all other lions—only a shade or so lighter, in fact, than those of their mother Tabby.

It's true that a white lion was born in a game park in Florida some years ago—but after a year it had turned into the normal tawny color. There's not the slightest sign of that happening with Temba and Tombi, who are now over two years old. In other words, they were probably the first genuinely white lions the world had ever seen. What's more, in October 1975, some nine months after the births of Temba and Tombi, *another* white cub, a female named Phuma—from the Zulu word for "to stand out," "to be out of the ordinary"— was born to another of the lionesses of the Machaton pride.

Chris McBride couldn't be absolutely sure this time of the identity of the mother—apart from the fact that as it is extremely unusual for a lion to mate with a female of the pride while she is still rearing her cubs, it's most unlikely that it was Tabby, the mother of Temba and Tombi.

It is, therefore, almost a certainty that the wayward gene that was responsible for the three white lions of Timbavati must have come from the father—from Agamemnon, that is, or possibly from Achilles if he really is Agamemnon's twin.

But in view of all the difficulties that beset any creature of the wild which stands out from the rest of its kind and doesn't melt into its natural background, have the white lions of Timbavati any future, either individually or as a new sub-species? That's a question to which Chris McBride has given a lot of thought:

⁂

From an evolutionary point of view, it might be said that the white lions are largely a curiosity. There are now white tigers in zoos in Delhi, Bristol and Washington, all bred from one white male tiger who was found in the wild, and was mated with his granddaughter and in turn with her offspring. The propagation of these white tigers has been a man-made situation. Theoretically, this would never occur in the wild on any long-term basis because it is not an advantage for a tiger—or a lion, for that matter—to be white.

On the other hand, the fact that there is a steady gradation from tawny to white among the Machaton pride seems to indicate that there is a gene here that tends to work toward a white strain.

Circuses and Zoos / 137

The white lions are beautiful. They were irresistible as cubs, and I am eager to find a way to ensure the survival of the strain.

Natural selection usually prevents brother and sister lions from mating with each other in the wild. So it is unlikely that Temba and Tombi would ever mate naturally, even if they were thrown out of the pride together. It could be stage-managed, of course, by putting them together in some sort of enclosure. But once you place them in such an artificial environment, there would be very little chance that they would ever be accepted back into the pride.

I've often wondered whether the best idea wouldn't be to hedge our bets and send the white male, Temba, off to a zoo now, quickly, before he is thrown out of the pride and exposed to the dangers outside the protection of the reserve; and while he is still young enough to settle down in a zoo. If we did that, it would provide revenue for the reserve for use in its wildlife management program. The funds could also be used to purchase electronic tracking and monitoring equipment to enable us to keep in constant touch with Tombi and Phuma.

In this way we would have the best of two worlds. Breeding experiments could be carried out with Temba in a zoo, under controlled conditions. And, at the same time, we would still have two white lions in the wild to continue to study.

If Tombi is not thrown out by the pride, there would be a strong possibility of her mating with Agamemnon, and there might well be further white cubs. In the meantime Phuma would also be growing up in the wild, and we would have another chance because she too might mate with Agamemnon, or even with some other pride male—possibly Vela—who could well carry the white gene.

In this way we would at least be doing everything in our power to ensure that this is not the end, but only the beginning of the story of the white lions of Timbavati.

From: *The White Lions of Timbavati*, by Chris McBride.

The Zoo

Johnny Morris

A keeper at a zoo has unrivalled opportunities, of course, of getting to know the animals in his charge—especially if he is someone like Johnny Morris whose tales are as young at heart as the children who visit zoos. Here are two of his stories. In the first, "Keeper" Morris visits the zoo at Copenhagen:

In the Gorilla Kitchen

One of the things that people always ask at the Zoo is "What time is feeding time?" We all find it interesting to watch animals feed. Needless to say the animals find feeding time very interesting too. And of course feeding is very important. It is something that we all have to do to keep alive. But one of the main things about feeding, as we know, is that we feel better and more contented if we eat the foods that we like. And so we try to eat the things we like and the more different things we can find the better we like it.

If you came with me into the Gorilla Kitchen at the Zoo you would find lots of different things that gorillas like. I call it the Gorilla Kitchen but actually it's the kitchen where all the apes' food is prepared.

Now the gorilla "Congo" lives right next to the Gorilla Kitchen. A thick wooden door leads from the kitchen into Congo's living-room, and sometimes when he is a bit impatient, or I am being rather slow getting his food ready, he thumps the door with his fist. It's a terrifying noise; it sounds like a big field-gun going off. And so I hurry up with his food.

Gorillas are of course vegetarians, and so I get ready for Congo half a dozen oranges, half a dozen apples, a bunch of bananas, a small loaf of brown bread, a large bunch of spring onions, a bunch of carrots and a cabbage, and he is very fond of a nice big slice of gorilla pudding.

Gorilla pudding looks rather like cold bread pudding and it's made in much the same way. Anyway, Congo is very fond of gorilla pudding.

The way that we give Congo his food is quite interesting. We don't give it to him all at once. This is how it's done. On the wall that separates the Gorilla Kitchen from Congo's living-room is a black steel box. It's cemented firmly into the wall. It is like a great big letterbox—like the ones that you sometimes see on the back of a door and you have to lift the lid to take the letters out.

All I have to do is drop six oranges into the box from my side, and Congo puts his big black hand through the large slot on his side and pulls them out. I post the oranges to him and he sort of unposts them and takes them away to eat. He can quite easily carry four or five big oranges in one hand.

I often watch him take his oranges away and sit on his big tree-trunk to eat them. He looks as though he is sampling the cooking of a world-famous chef; I can see him through a small peep-hole in the steel letter-box. When he's finished he comes back to the letter-box to see what else I've posted for him.

Sometimes I keep him waiting for just a minute so that I can look into his big black eyes. There he is peering through the slot in the letter-box. His tremendous face is only an inch or two from mine. He knows I am there because he can see my eye at the peep-hole, and he's puzzled and a little cross that I have not followed up the oranges with the spring onions and gorilla pudding. I put in a bunch of spring onions and he takes them away to his tree-trunk, eats them and comes back to the letter-box. He breathes heavily through the slot, and I must say it smells very oniony; so I pop in a bunch of carrots. Congo is very fond of carrots.

He takes them to the tree-trunk and chooses the biggest one of the bunch, smells it and then sort of smokes it like a cigar. He rolls it from one corner of his mouth to the other, just like a very tough gangster trying to work out how to rob the Bank of England. There he sits, this enormous gorilla smoking a carrot and thinking hard, thinking very hard indeed.

Or perhaps he's wondering why his regular keeper no longer goes in to play with him. His keeper used to do that until Congo got a little difficult. And you can easily understand why. Congo enjoyed playing with his keeper; he was very well-behaved and gentle, and never very rough. He wouldn't dream of hurting his keeper. But like all of us when the playtime came to an end, he didn't like it and he didn't want his keeper to leave him. When he was a very young gorilla the keeper could always get away fairly easily, but as Congo grew bigger and bigger this became more and more difficult. Congo couldn't understand that the keeper had important things to do other than play with him; and when the time came for the keeper to go, Congo would just take him firmly by the arm and look at him seriously as if to say "Now come, you're my friend, my very best friend, you're not going to leave me, are you?"

Congo would do anything to prevent the keeper leaving, even if it meant pulling him to pieces. And he could easily do that. So the keeper can no longer go in

with him. You cannot argue with a gorilla who does not want you to leave. At the moment Congo weighs about 37 stone (512 lbs.) and he is immensely powerful. It is difficult to say just how powerful. But here is an example of his incredible strength.

There was a small window in the wall of his living-room; it was only about a foot square and the glass was very special glass, an inch thick and extremely tough. Congo often used to look through this window —it gave him a pleasant view; until one day, for some unknown reason, he got a little angry with the glass and smashed it. It smashed in a shower of pieces!

Johnny Morris then recounted his meeting with some enthusiastic lemurs in Copenhagen Zoo:

Not-So-Demure Lemurs

You don't very often see a ring-tail lemur; they are found only in Madagascar and so they are comparatively rare. Their size is about halfway between a squirrel and a cat, their fur is soft pale grey and their eyes are very round, wide and brown, and full of wonderment. They seem to be astonished at everything they see. You can almost hear them saying "Cor, just look at all those people out there, what funny-looking things they are staring in at us staring out." But it is of course their tails that are so beautiful.

Their tails are nearly twice the length of their bodies and they are silver grey with black rings and delicately feathery. At the Zoo they have a big bare tree to play about in and sit in, and to see eighteen of them sitting in this twisted tree with their black ringed tails all hanging down is really one of the prettiest sights in the Zoo. Their tails look like the sallies on bell ropes in a church tower. You feel that if you gave them a gentle tug a little bell would tinkle somewhere.

When lemurs walk about they carry their tails high in the air like gay black-and-white banners, and when they feel sleepy they often curl their tails around their necks like peppermint fur collars. If ever you get a chance to see a ring-tail lemur go and see it. You couldn't wish for a more attractive animal.

I had never handled a ring-tail lemur, and so I asked the director of the Copenhagen Zoo if I might go into their enclosure. He was a jolly friendly man and he said "But of course you can go in with them, take some grapes and a little biscuit and some chopped apple and a handful of willow leaves, they love willow leaves. But you have to be just a little bit careful though, because one of them is inclined to bite a little." And I said "Oh I see, thank you very much."

And so I took a dish of grapes, biscuits and chopped apple, and a handful of willow leaves, and the keeper let me into their enclosure, and at once the lemurs came running and leaping at me.

There are so many nice things about lemurs. Once they are tame they are not a bit shy and not a bit suspicious. They accept you as a friend at once. They are sure that you will do them no harm, and they leap and scamper about all over you. I was absolutely loaded with ring-tail lemurs. Their little round brown eyes were staring with astonishment at my face, and they were looking to see if I'd washed behind my ears, opening my mouth, inspecting my teeth, looking down my collar, studying the label on my tie, and one of them spent about five minutes trying to read the word "Keeper" written on my hat. He just couldn't make it out. And then suddenly somebody nipped my ear and I shouted.

I shouldn't have shouted out quite so loud because it frightened the lemurs a lot. They leapt away from me and all huddled in a corner. They were absolutely flabbergasted that I should shout out like that. Eighteen lemurs sat and stared at me. Thirty-six round brown eyes wondering what on earth had made me make such a fuss. "He really shouldn't make a noise like that," they seemed to be saying. "After all we trusted him absolutely, and to make such a short sharp horrible noise just when we were climbing all over him and making friends was really just a bit too much."

However, after some moments of reproachful staring, one by one the lemurs came scampering back to me, and once again I was loaded with lemurs. And this time I watched carefully out of the corner of my eye, and sure enough the little lemur on my left shoulder very gently took the lobe of my ear and put it in his mouth, and I imagine he was about to bite when I said to him "Have a willow leaf." He didn't bite me; he took the willow leaf. I expect that was what he wanted in the first place, but he didn't know how to ask for it. After all the director did not say that one of them did bite, he definitely said that one of them is inclined to bite. And there is a difference, isn't there?

From: *Animal Magic*, by Johnny Morris.

News of Gnus

Gerald Durrell

It is fitting that our closing section should be by Gerald Durrell, for no one has collected animals more assiduously, or cared for them more devotedly. Although he now has a zoo of his own on Jersey, in the Channel Islands, back in 1945 he was working as a student keeper at Whipsnade. Among the most entrancing of the animals there, he found, were the gnus.

All the gnus are pretty unbelievable antelopes to look at, but the white-tailed has a particularly heraldic and mythical look about it. The head is blunt and the muzzle broad; the horns, curved like hunting horns, sweep down low over the eyes before curling upwards into sharp points so the animal is forced to peer under them in a myopic manner; a white beard juts out under the chin and another tuft of bristles decorates the top of the muzzle; the white mane is thick—a forest of uncombed tufts and sprigs; and a great sporran of hair grows between the forelegs. The long, sweeping, silky white tail is their best feature and they use it with all the elegance of an Oriental dancer with a scarf. Combined with this extraordinary appearance (which makes them look as though they have been made up out of bits of several different animals) are the gnus' extraordinary movements and the posturing they indulge in at the slightest provocation. To watch

these idiotic creatures, prancing, gyrating and snorting, their tails curling up over their backs, was one of the funniest sights I have seen.

Their movements were so complex that it was difficult to fit them into a category. It could really only be described as something like an acute attack of St. Vitus's Dance. Some of it resembled folk dancing of sorts, but it seemed a little vigourous. The only folk dances I have ever witnessed were danced by elderly, aesthetic ladies with fringes and strings of beads and they were nothing like the gnus' wild jitterbugging. Certainly there was a suggestion of ballet about it—of the more energetic and sweat-provoking sort—but the movements were too unorthodox for even the most frantically modern ballerina. This dance—or disease—is well worth watching. When the curtain rises, as it were, the gnus are facing you, bunched together, frowning through a forest of tufts and sprigs of hair. One member of the troop assumes the leadership and he (or she) starts the dance by giving a purring snort of astonishing loudness, a sort of preliminary "now girls, all together." Then the whole lot mince a few steps on slender legs; they stand again, legs quivering, tails twitching almost in unison; then the leader gives another snort which invariably has the effect of making the whole troop lose their heads. Forgotten is the grouping and precision which delights the eye in ballet. With stamping, polished hooves, away they go, tails curled, bucking, kicking legs thrown out at ridiculous and completely unanatomical angles. The leader keeps up a barrage of frantic snorts—orders which no one obeys. Then, quite suddenly, they all stop and stare at you from under their horns in horrified disapproval at your unmannerly laughter.

It was, in fact, the white-tailed gnu's habit of dancing and its insatiable curiosity that led it to the brink of extermination. In the early days, during the colonisation of South Africa, the white-tailed gnu was found in thousands and the early Dutch settlers killed it relentlessly, first because its meat, dried into biltong, could be used as food instead of slaughtering valuable cattle and sheep, and, secondly, the quicker it was out of the way, they thought, the more grazing room there would be for domestic stock. So in a very short time what had been the most numerous of the African antelopes became one of the rarest. Its engaging curiosity, which would make a herd stand there peering at the hunters while they were shot down, was partially the cause of their downfall and also the endearing fact that they loved to perform their dances and would prance and waltz round wagons bristling with guns, thus forming one of the easiest of targets. Now the white-tailed gnu is no longer a truly wild animal. Just over two thousand are left in small parks and on private farms and a scant hundred specimens in the zoos of the world.

As I watched our gnus posturing, rampant on a field of green grass, I thought how dull the African scene must be without these gay, frenetic dancers of the veld. It seems that always progress destroys the happy and original, making everything banal, replacing these joyous prancing creatures with the dull, cud-chewing, utilitarian cow.

As well as the white-tailed gnu, we had a solitary brindled gnu, an animal much the same in shape though a bit thicker set and with a gingery fawn coat with chocolate brindles and a black mane and tail. If anything, this animal was even more mentally defective than the white-tailed gnu; his gyrations were even more wildly extravagant and his deep, belching roars of alarm rattled from deep in his chest like machine-gun fire. He was an incredibly nervous beast—more than liable, if you frightened him, to break a leg or do himself some other injury—so it was with feelings of acute apprehension that Harry and I received the news that Brinny had to be caught up and transported to London Zoo where they had acquired a mate for him.

Thursday dawned and the lorry arrived with a tall, narrow crate into which we had to try and cajole an exceedingly nervous, high-spirited and agile gnu. We had let Brinny out into his paddock for a brief airing that morning; I had then enticed him back into the double stable by bribing him with oats, and now we had him safely locked up in one of the loose-boxes. Next we had to manhandle the massive crate off the back of the lorry and get it into position facing the door of the empty loose-box, then raise the sliding door on the end of the crate. This took us some time and we were forced, not unnaturally, to make a fair bit of noise over it, which Brinny took grave exception to. He belched and snorted and reared and several times attempted to kick the side of the stable out. Once having got the crate in position, we went away for half an hour to discuss strategy and to let Brinny calm down a bit.

"Now, boy," said Harry, "this is what we'll do, see. I'll be atop the crate and 'andle the slide, but once I've got that slide up I can't see when 'e goes into the crate so you'll 'ave to tell me when to drop the slide, see? Now, I want you to take the ladder into the stable next door, then you take this bit of two-by-two, lean over the dividing wall and when 'e gets near the crate just give 'im a tap on the rump—only a tap, mind—that's all 'e'll need, just enough to make 'im run into the crate. Then, when 'e's in, you give a yell and I'll drop the slide, see?"

"You make it sound so simple," I said bitterly.

"Let's 'ope it is," said Harry, grinning.

We trooped back to the stable, where Brinny was still gurking fiercely, and I manoevered the ladder into the stable next door, took my piece of wood and climbed up and peered over the partition. Brinny stared up at me, horror-stricken that I should do such a dastardly thing as to take him in the rear. His mane and beard looked wild and uncombed and gave him the air of having just arisen, dishevelled, from his bed. His eyes rolled, his nostrils grew wide with every snort and his curved black horns glinted like knives as he pranced and gyrated within the confines of the stable.

"You ready, boy," shouted Harry from outside.

I eased my piece of wood over the partition and made sure of my foothold on the ladder.

"Okay!" I yelled. "Fire away."

Brinny, who had been staring up at me with the expression of a spinster who had at last actually found a man under her bed, naturally waltzed round to face the door and watch as the slide on the crate was slowly raised. He snorted like a volcano and minced from side to side nervously. As he was not watching me I manoevered my piece of wood into position. I grasped it firmly with one hand and cupped my other hand over the end. I could not have chosen a more unfortunate grasp.

"I'm going to chivy him, Harry," I called.

"All right, boy," said Harry.

I carefully lowered my piece of wood towards Brinny's rotund and quivering backside. As the end of my stick touched his glossy hide it was as though I had touched a match to a short fuse on a barrel of TNT. Everything seemed to happen at once. Feeling the stick, Brinny wasted no time, he leapt straight into the air and tried to kick his heels over his horns. He caught my piece of wood and shot it skywards like a rocket so that it crashed against the roof of the stable. My hand, which was cupped round the end of it, was therefore crushed against the ceiling as though caught by a pile-driver. The pain was so excruciating that I dropped the wood and tried to struggle back over the partition which I was half lying on. I could feel the ladder swaying under me. At that moment Brinny uttered a particularly prodigious snort, put his head down and rushed into the crate.

"Slide, Harry, slide!" I yelled desperately just as the ladder gave under me and I fell into the stable. The slide crashed into position and we had Brinny imprisoned—but only just, for he had galloped into the crate and hit the end of it with his horns, making the whole structure sway like a ship in a hurricane. Then he attacked the end of the crate with short jabs of his horns and splintered wood began to fly in all directions. People started running about in pursuit of hammer and nails to repair the damage before Brinny could force his way out. Harry, perched precariously on the swaying crate, peered down at me.

"You all right, boy?" he inquired anxiously.

I got up a trifle shakily; my hand felt as though it had been trodden on by an elephant and was already beginning to swell.

"I'm okay, but I think I've bust my hand," I said.

This was my first honourable wound in the course of duty, as it were.

From: *Beasts in My Belfry*, by Gerald Durrell.

ANIMALS AND MAN

The relationships between animals and humans have varied from age to age. Primitive man, before he settled down to cultivate the land, lived by hunting. The animals he hunted were so essential to his survival that he not only respected them but regarded them with religious awe, and his killing of them was accompanied by all kinds of rituals designed to ensure their continuing abundance, and to propitiate the unseen powers that had created or entered into them. Even when man became a herdsman rather than a hunter, his feelings towards his domestic animals often remained much the same—as indeed they still do among the surviving primitive peoples today, who depend upon their herds for their survival. The Dinkas of the Sudan, for example, make the slaughter of an animal a religious sacrifice and their poetry, songs and folklore are focused on their cattle.

These primitive feelings towards animals survived into the civilizations of the ancient world, where the sense of a bond between man and animal remained strong. In ancient Egypt a number of animals were regarded as divine representatives of the super-natural powers and the gods and goddesses were usually depicted as half-animal and half-human. In ancient Greek mythology, too, the gods and goddesses frequently assumed the forms of animals.

It was perhaps only with the evolution of the idea of one God, who had entered into a special relationship with man, that a superior attitude towards the "lower" creation came into being. Then man began to think of himself as the ruler over all the animals, and of his world as the central point of the universe, so that his attitude towards the creatures with whom he shared the earth became increasingly arrogant. There were, of course, many individual exceptions. In the Middle Ages the sense of man's kinship with the animals was never completely lost. Many of the Christian saints felt it was part of their religion, and had a specially close relationship with animals. St. Anthony preached to the fishes and the pigs. St. Roch declared he had found a greater faith in dogs than in men. St. Francis taught the birds and called them brothers. St. Hugh of Lincoln treasured his squirrel and his swan. It was never forgotten, either, that according to the New Testament story, the first living creatures to adore the Holy Child in the manger at Bethlehem were the beasts with whom He shared the stable. And there's the charming old legend of St. Anthony's mule which—when the saint held aloft the Communion bread and wine—left its oats and went down on its knees in adoration.

As time went by a whole series of blows shook man's confidence in himself as a unique creature. The realization that the world on which he and the animals lived was not, after all, the focal point of the universe but merely one of many bodies in space, was the first of them. Another discovery was the demonstration by the late eighteenth-century geologists and palaeontologists that the fossil record showed that the world and all the living forms within it did not come into being by one single act of creation at one specific date. The severest blow of all, of course, was the conclusion put forward in 1859 by Charles Darwin in his momentous book *On the Origins of Species by Means of Natural Selection*, namely that man, like all the other animals, had evolved from other forms of life in the far distant past. In our own time man has been further humbled by the discoveries of the astronomers and astro-physicists which reduce even more the importance of our own globe, and suggest that forms of life of a far higher quality than our own may possibly exist somewhere in the cosmos; and also by the realization that there are elements in man's nature which make him a most dangerous guardian of his own scientific discoveries and which may eventually lead him to blow himself and all other forms of life out of existence.

All these, and many other blows to man's pride, haven't put a complete stop to the slaughter and exploitation of animals for food, transport, greed, war or sport—but they have made possible the gradual growth of a new attitude towards animals and the natural environment, comparable in some respects to the attitude of ancient man. Perhaps man is now prepared to study the habits and capabilities of animals with a renewed respect, sense of kinship, and openness of mind. It seems appropriate to begin our look at the relationship of man to animals with Konrad Lorenz's fascinating study of the difference in attitude towards man that the two basic types of dog display.

Dogs and Man—the Covenant

Konrad Lorenz

Acclaimed naturalist Konrad Lorenz analyzes the rare friendship of men and dogs in this selection from his classic study of animal behavior, King Solomon's Ring.

There is no faith which has never yet been broken, except that of a truly faithful dog. Of all dogs which I have hitherto known, the most faithful are those in whose veins flows, beside that of the golden jackal *(Canis aureus)*, a considerable stream of wolf's blood. The northern wolf *(Canis lupus)* only figures in the ancestry of our present dog breeds through having been crossed with already domesticated Aureus dogs. Contrary to the widespread opinion that the wolf plays an essential role in the ancestry of the larger dog breeds, comparative research in behaviour has revealed the fact that all European dogs, including the largest ones such as Great Danes and wolfhounds, are pure Aureus and contain, at the most, a minute amount of wolf's blood. The purest wolf-dogs that exist are certain breeds of Arctic America, particularly the so-called malamutes and huskies. The Eskimo dogs of Greenland also show only slight traces of Lapland Aureus characters, whereas the Arctic breeds of the Old World, such as Lapland dogs, Russian lajkas, samoyeds and chows, certainly have more

Aureus in their constitution. Nevertheless the latter breeds derive their character from the Lupus side of their ancestry and they all exhibit the high cheekbones, the slanting eyes and the slightly upward tilt of the nose which give its specific expression to the face of the wolf. On the other hand the chow, in particular, bears the stamp of his share of Aureus blood in the flaming red of his magnificent coat.

The "sealing of the bond," the final attachment of the dog to one master, is quite enigmatical. It takes place quite suddenly, within a few days, particularly in the case of puppies that come from a breeding kennel. The "susceptible period" for this most important occurrence in the whole of a dog's life is, in Aureus dogs, between 8 and 18 months, and in Lupus dogs round about the sixth month.

The really single-hearted devotion of a dog to its master has two quite different sources. On the one side, it is nothing other than the submissive attachment which every wild dog shows towards its pack leader, and which is transferred, without any considerable alteration in character, by the domestic dog to a human being. To this is added, in the more highly domesticated dogs, quite another form of affection. Many of the characteristics in which domestic animals differ from their wild ancestral form arise by virtue of the fact that properties of body structure and behaviour, which in the wild prototype are only marked by some transient stages of youth, are kept permanently by the domestic form. In dogs, short hair, curly tail, hanging ears, domed skull and the short muzzle of many domestic breeds are features of this type. In behaviour, one of these juvenile characters which has become permanent in the domestic dog, expresses itself in the peculiar form of its attachment. The ardent affection which wild canine youngsters show for their mother and which disappears completely after they have reached maturity, is preserved as a permanent mental trait of all highly domesticated dogs. What originally was love for the mother is transformed into love for the human master.

Thus the pack loyalty, in itself unaltered but merely transferred to man, and the permanent childlike dependency resulting from domestication, are two more or less independent springs of canine affection. One essential difference in the character of Lupus and Aureus dogs is attributable to the fact that these two springs flow with different strength in the two types. In the life of a wolf, the community of the pack plays a vastly more important role than in that of a jackal. While the latter is essentially a solitary hunter and confines itself to a limited territory, the wolf packs roam far and wide through the forests of the North as a sworn and very exclusive band which sticks together through thick and thin and whose members will defend each other to the very death. That the wolves of a pack will devour each other, as is frequently asserted, I have strong reason to doubt, since sled-dogs will not do so at any price, even when at the point of starvation, and this social inhibition has certainly not been instilled into them by man.

The reticent exclusiveness and the mutual defense at all costs are properties of the wolf which influence favourably the character of all strongly wolf-blooded dog breeds and distinguish them to their advantage from Aureus dogs, which are mostly "hail-fellow-well-met" with every man and will follow anyone who holds the other end of the lead in his hand. A Lupus dog, on the contrary, who has once sworn allegiance to a certain man, is forever a one-man dog and no stranger can win from him so much as a single wag of his bushy tail. Nobody who has once possessed the one-man love of a Lupus dog will ever be content with one of pure Aureus blood. Unfortunately this fine characteristic of the Lupus dog has against it various disadvantages which are indeed the immediate results of the one-man loyalty. That a mature Lupus dog can never become your dog, is a matter of course. Worse, if he is already

yours and you are forced to leave him, the animal literally becomes mentally unbalanced, obeys neither your wife nor children, sinks morally, in his grief, to the level of an ownerless street cur, loses his restraint from killing and, committing misdeed upon misdeed, ravages the surrounding district.

Besides this a predominantly Lupus-blooded dog is, in spite of his boundless loyalty and affection, never quite sufficiently submissive. He is ready to die for you, but not to obey you; at least, I have never been able to extract implicit obedience from one of these dogs—perhaps a better dog trainer than I might be more successful. For this reason, it is seldom that you see, in a town, a chow without a lead walking close beside his master. If you walk with a Lupus dog in the woods, you can never make him stay near you. All he will do is to keep in very loose contact with you and honour you with his companionship only now and again.

Not so the Aureus dog. In him as a result of his age-old domestication, that infantile affection has persisted which makes him a manageable and tractable companion. Instead of the proud loyalty of the Lupus dog, which is far removed from obedience, the Aureus dog will always grant you that servitude which, day and night, by the hour and by the minute, awaits your command and even your slightest wish. When you take him for a walk, an Aureus dog of a more highly domesticated breed will, without previous training, always run with you, keeping the same radius whether he runs before, behind or beside you and adapting his speed to yours. He is naturally obedient: he answers to his name not only when he wishes to and when you cajole him, but also because he knows that he must come. The harder you shout, the more surely he will come, whereas a Lupus dog, in this case, comes not at all but seeks to appease you from a distance with friendly gestures.

Opposed to these good and congenial properties of the Aureus dog are unfortunately some others which also arise from the permanent infantility of these animals and are less agreeable for an owner. Since young dogs under a certain age are, for members of their own species, "taboo"—that is, they must not under any circumstances be bitten—such big babies are often correspondingly trustful and importunate towards everybody. Like many spoilt human children who call every grown-up "uncle," they pester people and animals alike with overtures to play. If this youthful property persists, to any appreciable extent, in the adult domestic dog, there arises a very unpleasant canine character, or rather the complete lack of such a commodity. The worst part of it lies in the literally "dog-like" submission that these animals, who see in every man an "uncle," show towards anyone who treats them with the least sign of severity; the playful storm of affection is immediately transformed into a cringing state of humility. Everyone is acquainted with this kind of dog which knows no happy medium between perpetual, exasperating "jumping up," and fawningly turning upon its back, its paws waving in supplication. You shout, at the risk of offending your hostess, at the infuriating creature that is trampling all over your person and covering you from head to foot with hairs. Thereupon the dog falls beseechingly upon his back. You speak kindly to him, to conciliate your hostess and, quickly leaping up, the brute has licked you right across the face and now continues unremittingly to bestrew your trousers with hairs.

Yet my affections do not belong entirely to Lupus dogs, as the reader might conclude from this little canine characterology. No Lupus-blooded dog has offered his master such unquestioning obedience as our incomparable Alsatian—an Aureus dog. But both sets of qualities can be combined. It would, of course, be quite impossible for the dog-breeder to make the predominantly Lupus dog catch up, in one stride, with the Aureus dog which has been domesticated for a few thousand years longer, but there is another way.

Some years ago my wife and I each possessed a dog, I an Alsatian bitch, Tito, my wife a little chow bitch, Pygi. A son of Tito's, Booby by name, married the chow bitch Pygi. This happened quite against the will of my wife who, naturally enough, wanted to breed pure chows. But here we discovered, as an unexpected hindrance, a new property of Lupus dogs: the monogamous fidelity of the bitch to a certain dog. My wife travelled with her bitch to nearly all the chow dogs in Vienna, in the hope that one at least would find Pygi's favour. In vain—Pygi snapped furiously at all her suitors; she only wanted her Booby and she got him in the end, or rather he got her by reducing a thick wooden door, behind which Pygi was confined, to its primary elements. And therewith began our crossbred stud of chow and Alsatian.

At the moment, our breed contains very little Alsatian blood, because my wife, during my absence in the war, twice crossed in pure chows; this was inevitable, for otherwise we should have been dependent on inbreeding. As it is, the inheritance of Tito shows itself clearly in the psychological respect, for the dogs are far more affectionate and much easier to train than pure-blooded chows, although, from an external point of view, only a very expert eye can detect the element of Alsatian blood. I intend to develop further this mixed breed, and to continue with my plan to evolve a dog of ideal character.

From: *King Solomon's Ring*, by Konrad Lorenz.

Chimpanzee Society

Jane Goodall

It is now generally accepted that the study of animal societies can teach us a good deal about our own. One of the most interesting and important of such studies in recent years was that carried out by Jane Goodall (who became the wife of another naturalist, Hugo van Lawick) among a community of chimpanzees in the Gombe Reserve on the shores of Lake Tanganyika. Its outstanding feature was the combination of exact observation, scientific objectivity, and affection for the animals themselves. Here she tells us the story of Mike, and his rise in chimpanzee society:

Mike's rise to the number one or top-ranking position in the chimpanzee community was both interesting and spectacular. In 1963 Mike had ranked almost bottom in the adult male dominance hierarchy. He had been the last to gain access to bananas, and had been threatened and actually attacked by almost every other adult male. Indeed, at one time he had appeared almost bald from losing so many handfuls of hair during aggressive incidents with his fellow apes.

When Hugo and I had left the Gombe Stream at the end of that year, prior to getting married, Mike's position had not changed; yet when we returned, four

months later, we found a very different Mike. Kris and Dominic told us the beginning of his story—how he had started to use empty four-gallon paraffin cans more and more often during his charging displays. We did not have to wait many days before we witnessed Mike's techniques for ourselves.

There was one incident that I remember particularly vividly. A group of five adult males, including top-ranking Goliath, David Greybeard and the huge Rudolf, were grooming each other—the session had been going on for some twenty minutes. Mike was sitting on his own about thirty yards from them, frequently staring towards the group, occasionally idly grooming himself. All at once Mike calmly walked over to our tent and took hold of an empty paraffin can by the handle. Then he picked up a second can and, walking upright, returned to the place where he had been sitting before. Armed with his two cans, Mike continued to stare towards the other males and, after a few minutes, he began to rock from side to side. At first the movement was almost imperceptible, but Hugo and I were watching him closely. Gradually he rocked more vigourously, his hair slowly began to stand erect, and then, softly at first, he started a series of hoots. As he called, Mike got to his feet and suddenly he was off, charging towards the group of males, hitting the two cans ahead of him. The cans, together with Mike's crescendo of hooting, made the most appalling racket: no wonder the erstwhile peaceful males rushed out of the way. Mike and his cans vanished down a track and, after a few moments, there was silence. Some of the males reassembled and resumed their interrupted grooming session, but the others stood around somewhat apprehensively.

After a short interval that low-pitched hooting began again, followed, almost immediately, by the appearance of the two rackety cans with Mike close behind them. Straight for the other males he charged, and once more they fled. This time, even before the group could reassemble, Mike set off again: but he made straight for Goliath—and even he hastened out of Mike's way like all the others. Then Mike stopped and sat, all his hair on end and breathing hard. His eyes glared ahead and his lower lip was hanging slightly down so that the pink inside showed brightly and gave him a wild appearance. Rudolf was the first of the males to approach Mike, uttering soft pant-grunts of submission, crouching low and pressing his lips to Mike's thigh. Then he began to groom Mike and two other males approached, pant-grunting, and began to groom him also. Finally David Greybeard went over to Mike, laid one hand on his groin, and joined in the grooming. Only Goliath kept away, sitting on his own and staring towards Mike. It was obvious that Mike constituted a serious threat to Goliath's hitherto unchallenged supremacy.

Mike's deliberate use of man-made objects was probably an indication of superior intelligence. Many of the adult males had, at some time or another, dragged a paraffin can to enhance their charging displays, in place of the more normal branches or rocks; but only Mike apparently had been able to profit from the chance experience and learned to seek out the cans deliberately to his own advantage. The cans, of course, made a great deal more noise than a branch when dragged along the ground at speed, and, after a while, Mike was actually able to keep three cans ahead of him at once for about sixty yards as he ran flat out across the camp clearing. No wonder that males, previously his superiors, rushed out of Mike's way.

Charging displays usually occur when a chimpanzee becomes emotionally excited; when he arrives at a food source, joins up with another group or when he is frustrated. But it seemed that Mike actually planned his charging displays—almost, one might say, in cold blood. Often, when he got up to fetch his cans, he showed no visible signs of frustration or excitement—that came afterwards when, armed with his display props, he began to rock from side to side, raise his hair, and hoot.

Eventually Mike's use of paraffin cans became dangerous for he learned to hurl them ahead of him at the close of a charge—once he got me on the back of my head, and once he hit Hugo's precious film camera. We decided to remove all the cans, and, for a while, went through a nightmare period since Mike tried to drag about all manner of other objects. Once he got hold of Hugo's tripod—luckily when the camera was not mounted—and once he managed to grab and pull down the large cupboard in which we kept a good deal of food and all our crockery and cutlery. The noise and the trail of destruction were unbelievable. Finally, however, we managed to dig things into the ground or hide them away, and Mike had to resort to branches and rocks like his companions. By that time, however, his top-ranking status was assured, although it was fully another year before Mike himself seemed to feel quite secure in his position. He continued to display very frequently and vigourously, and the lower-ranking chimps had increasing reason to fear him for often he would attack a female or youngster viciously at the slightest provocation. In particular, as might be expected, a tense relationship prevailed between Mike and the ex-dominant male, Goliath.

Goliath did not relinquish his position without a struggle. His displays also increased in frequency and vigour and he, too, became more aggressive. Indeed, there was a time, towards the start of this battle for dominance, when Hugo and I feared for Goliath's sanity. After attacking a couple of youngsters and charging back and forth dragging huge branches, he would sit, his hair on end, his sides heaving from exertion, a froth of saliva glistening at his half-open mouth, and a glint in his eyes that, to us, looked not far from madness. We actually had a weld-mesh iron cage built in Kigoma and, when this had been set up in camp, we retreated inside when Goliath's temper was at its worst.

One day, when Mike was sitting in camp, a series of distinctive, rather melodious, pant-hoots, with characteristic quavers at the close, announced the return of Goliath who, for two weeks, had been somewhere down in the south of the Reserve. Mike responded immediately, hooting himself and charging across the clearing. Then he climbed a tree and sat staring over the valley, every hair on end. A few minutes later Goliath appeared and, as he reached the outskirts of the camp clearing, he commenced one of his spectacular displays. He must have seen Mike for he headed straight for him, dragging a huge branch. Then he leapt up into a tree near that of Mike and was still.

For a moment Mike stared towards him and then he too began to display, swaying the branches of his tree, swinging to the ground, hurling a few rocks and, finally, climbing up into Goliath's tree and swaying the branches there. When he stopped, Goliath immediately reciprocated, swinging about in the tree and rocking the branches. Presently, as one of his wild leaps took him quite close to Mike, Mike too displayed, and for a few unbelievable moments both of the splendid male chimpanzees were swaying branches within a few feet of each other until I thought the whole tree must crash to the ground. But an instant later both chimps were on the ground, displaying in the undergrowth. Finally they stopped and sat, staring at each other. It was Goliath who moved next, standing upright as he rocked a sapling; when he paused Mike charged past him, hurling a rock and drumming, with his feet, on the trunk of a tree.

This went on for nearly half an hour: first one male and then the other displayed, and each performance seemed to be more vigourous, more spectacular, than that preceding it. Yet during all this time, apart from occasionally hitting one another with the ends of the branches they swayed, neither chimpanzee actually attacked the other. Suddenly, after an extra long pause, it seemed that Goliath's nerve broke. He rushed up to Mike, crouched beside him with loud, nervous pant-grunts, and began to groom him with feverish intensity. For a few moments Mike ignored Goliath completely: then he turned and, with a vigour almost matching that of Goliath, began to groom his vanquished rival. And there they sat, grooming each other without pause, for over an hour.

That was the last real duel between the two males. From then on it seemed that Goliath accepted Mike's superiority, and a strangely intense relationship grew up between the two. They often greeted one another with much display of emotion, embracing or patting one another, kissing each other in the neck after which they usually started grooming each other. During these grooming sessions it appeared that the tension between them was eased, soothed by the close, friendly physical contact. Afterwards they sometimes fed or rested quite close to each other, looking peaceful and relaxed as though the bitter rivalry of the past had never been.

But it was with another chimpanzee, named David Greybeard, that Jane Goodall had the strongest ties of affection:

For me, of course, the saddest loss was when David Greybeard died. For David was the first chimpanzee to accept my presence and permit me to approach him closely. David, with his gentle disposition, who permitted a strange white ape to touch him. To me it represented a triumph of the sort of relationship which man can establish with a wild creature. Indeed, when I was with David Greybeard I sometimes felt that our relationship came closer to friendship than I would have thought possible with a completely free wild creature, a creature who had never known captivity.

In those early days I spent many days alone with David. Hour after hour I followed him through the forests, sitting and watching him whilst he fed or rested, struggling to keep up when he moved through a tangle of vines. Sometimes, I am sure, he waited for me—just as he would wait for Goliath or William. For when I emerged, panting and torn from a mass of thorny undergrowth, I often found him sitting, looking back in my direction. When I had emerged, then he got up and plodded on again. One day, as I sat near him at the bank of a tiny trickle of crystal-clear water, I saw a ripe palm nut lying on the ground. I picked it up and held it out to him on my open palm. He turned his head away. But when I moved my hand a little closer he looked at it, and then at me, and then he took the fruit and, at the same time, he held my hand firmly and gently with his own. As I sat, motionless, he released my hand, looked down at the nut, and dropped it to the ground. At that moment there was no need of any scientific knowledge to understand his communication of reassurance. The soft pressure of his fingers spoke to me not through my intellect but through a more primitive emotional channel: the separate evolution of man and chimpanzee was, for those few seconds, broken down. It was a reward far beyond my greatest hopes.

From: *In the Shadow of Man*, by Jane Goodall.

Man and Baboon

Desmond Morris

One of the recent invigorating shocks to our complacency was Dr. Desmond Morris's book, The Human Zoo, *in which he argues that many of the drives and aspirations of human beings are very similar to those of animals. He lists, for example, ten "golden rules" which apply equally to men and baboons in their struggle to reach the top. Here are two:*

1. *You must clearly display the trappings, postures and gestures of dominance.*

For the baboon this means a sleek, beautifully. groomed, luxuriant coat of hair; a calm, relaxed posture when not engaged in disputes; a deliberate and purposeful gait when active. There must be no outward signs of anxiety, indecision or hesitancy.

With a few superficial modifications, the same holds true for the human leader. The luxuriant coat of fur becomes the rich and elaborate costume of the ruler, dramatically excelling those of his subordinates. He assumes postures unique to his dominant role. When he is relaxing, he may recline or sit, while others must stand until given permission to follow suit. This is also typical of the dominant baboon, who may sprawl out lazily while his anxious subordinates hold themselves in more alert postures nearby. The situation changes once the leader stirs into aggressive action and begins to assert himself. Then, be he baboon or prince, he must rise into a more impressive position than that of his followers. He must literally rise above them, matching his psychological status with his physical posture. For the baboon boss this is easy: a dominant monkey is nearly always much larger than his underlings. He has only to hold himself erect and his greater body size does the rest. The situation is enhanced by cringing and crouching on the part of his more fearful subordinates. For the human leader, artificial aids may be necessary. He can magnify his size by wearing large cloaks or tall headgear. His height can be increased by mounting a throne, a platform, an animal, or a vehicle of some kind, or by being carried aloft by his followers. The crouching of the weaker baboons becomes stylized in various ways: subordinate humans lower their height by bowing, curtsying, kneeling, kowtowing, salaaming or prostrating.

The ingenuity of our species permits the human leader to have it both ways. By sitting on a throne on a raised platform, he can enjoy both the relaxed position of the passive dominant *and* the heightened position of the active dominant at one and the same time, thus providing himself with a doubly powerful display posture.

The dignified displays of leadership that the human animal shares with the baboon are still with us in many forms today. They can be seen in their most primitive and obvious conditions in generals, judges, high priests and surviving royalty. They tend to be more limited to special occasions than they once were, but when they do occur they are as ostentatious as ever. Not even the most learned academics are immune to the demands of pomp and finery on their more ceremonial occasions.

Where emperors have given way to elected presidents and prime ministers, personal dominance displays have, however, become less overt. There has been a shift of emphasis in the role of leadership. The new-style leader is a servant of the people who happens to be dominant, rather than a dominator of the people who also serves them. He underlines his acceptance of this situation by wearing a comparatively drab costume, but this is only a trick. It is a minor dishonesty that he can afford, to make him seem more "one of the crowd," but he dare not carry it too far or, before he knows it, he really will have become one of the crowd again. So, in other, less blatantly personal ways, he must continue to perform the outward display of his dominance. With all the complexities of the modern urban environment at his disposal, this is not difficult. The loss of grandeur in his dress can be compensated for by the elaborate and exclusive nature of the rooms in which he rules and the buildings in which he lives and works. He can retain ostentation in the way he travels, with motorcades, out-riders and personal planes. He can continue to surround himself with a large group of "professional subordinates"—aides, secretaries, servants, personal assistants, bodyguards, attendants, and the rest—part of whose job is merely

to be seen to be servile towards him, thereby adding to his image of social superiority. His postures, movements and gestures of dominance can be retained unmodified. Because the power signals they transmit are so basic to the human species, they are accepted unconsciously and can therefore escape restriction. His movements and gestures are calm and relaxed, or firm and deliberate. (When did you last see a president or a prime minister running, except when taking voluntary exercise?) In conversation he uses his eyes like weapons, delivering a fixed stare at moments when subordinates would be politely averting their gaze, and turning his head away at moments when subordinates would be watching intently. He does not scrabble, twitch, fidget or falter. These are essentially the reactions of subordinates. If the leader performs them there is something seriously wrong with him in his role as the dominant member of the group.

2. *In moments of active rivalry you must threaten your subordinates aggressively.*

At the slightest sign of any challenge from a subordinate baboon, the group leader immediately responds with an impressive display of threatening behavior. There is a whole range of threat displays available, varying from those motivated by a lot of aggression tinged with a little fear to those motivated by a lot of fear and only a little aggression. The latter—the "scared threats" of weak-but-hostile individuals—are never shown by a dominant animal unless his leadership is tottering. When his position is secure he shows only the most aggressive threat displays. He can be so secure that all he needs to do is to indicate that he is about to threaten, without actually bothering to carry it through. A mere jerk of his massive head in the direction of the unruly subordinate may be sufficient to subdue the inferior individual. These actions are called "intention movements," and they operate in precisely the same way in the human species. A powerful human leader, irritated by the actions of a subordinate, need only jerk his head in the latter's direction and fix him with a hard stare, to assert his dominance successfully. If he has to raise his voice or repeat an order, his dominance is slightly less secure, and he will, on eventually regaining control, have to re-establish his status by administering a rebuke or a symbolic punishment of some kind.

The act of raising his voice, or raging, is only a weak

sign in a leader when it occurs as a reaction to an immediate threat. It may also be used spontaneously or deliberately by a strong ruler as a general device for reaffirming his position. A dominant baboon may behave in the same way, suddenly charging at his subordinates and terrorizing them, reminding them of his powers. It enables him to chalk up a few points, and after that he can more easily get his own way with the merest nod of his head. Human leaders perform in this manner from time to time, issuing stern edicts, making lightning inspections, or haranguing the group with vigorous speeches. If you are a leader, it is dangerous to remain silent, unseen or unfelt for too long. If natural circumstances do not prompt a show of power, the circumstances must be invented that do. It is not enough to have power, one must be observed to have power. Therein lies the value of spontaneous threat displays.

From: *The Human Zoo*, by Desmond Morris.

Animals Can Be Almost Human

Max Eastman

In this new climate of opinion the tendency is to empha-size the similarities between humans and animals rather than the differences.

Hardly a week goes by that newspapers do not contain one or more stories about fabulous feats performed by animals. I have admired my animal neighbors all my life, but I confess that I find many of these stories a little too fabulous, others too sentimental, to suit me. I do not want animals to be supernatural; I want them to be natural. I am not half so much interested in tales of the intellectual prowess or moral heroism of some dog or cat or elephant as in learning about those traits of animals in general that are similar to ours and give us a sense of the kinship of life.

Take the giving of gifts and love tokens. According to Edward Armstrong, author of *The Way Birds Live*, even the custom of "saying it with flowers" is to be found among certain birds and insects. The male empid fly, for instance, wraps up a flower petal or a bit of food in a web of fine silk that he weaves with his front feet, and presents it to his bride. "Starlings," says Professor Armstrong, "carry flowers into their nesting-hole when the female is on the nest. A herring gull will pick up a shell or pluck a sea pink and, with great courtesy, lay it before the brooding mate."

Other birds come so near to being human that they express their sentimental emotions by talking baby talk—a trick I cannot endorse in either species. According to Konrad Lorenz, an outstanding naturalist, "Every delicacy the male jackdaw finds is given to the bride, and she accepts it with the plaintive notes typical of baby birds. The love whispers of the couple consist chiefly of infantile sounds."

A more dignified example of similarity between humans and animals is the ceremony of betrothal. Long engagements always seemed to me an unbiological affliction that man in a state of puritanical super-civilization has imposed upon himself. Among robins, however, extended engagements are an all but inflexible rule. They pair up in late December or January, but do not mate or start housekeeping until the end of March. Among jackdaws and wild geese betrothal occurs in the spring following birth, although neither species becomes sexually mature until a year later. Indeed, nearly all birds that marry for life are betrothed before they marry.

Another social custom commonly regarded as peculiarly human is the division of society into castes or classes, with the special privilege, oppression, cruelty and snobbery that go with it. You can see it in the hen-run, where a definite social hierarchy, or "pecking order," is always established. Every bird has a wholesome fear of those above her in rank and also knows which ones are below her. It is not always by tests of strength that this order is established; energy, nerve and, above all, self-assurance, also play a vital role. Just as among men, this hierarchy of status and prestige is liable to culminate in dictatorship.

An equally human aspect of this ladder of prestige is the snobbery it entails. Dr. Lorenz describes how a jackdaw of high rank fell in love with a young female among the lower orders. Within a few days the entire colony knew that this little low-class upstart, whom 80 per cent of them had been maltreating, could no longer receive a black look from anybody. She knew it, too, and made the fullest use of it. "She lacked entirely," Lorenz mourns, "that noble tolerance which jackdaws of high rank should exhibit towards their inferiors. She used every opportunity to snub former superiors. In short, she conducted herself with the utmost vulgarity."

Dr. Lorenz warns us against the sentimental notion that animals are morally "better" or "worse" than man. Moral judgements, he insists, are irrelevant where life is instinctive. In explaining the all-too-human sins of animals, he relates the sad tale of the alienation of a swan husband's tender affections by a determined female. Swans are monogamous, and supposedly faithful to their mates for life. But one old male swan "furiously expelled a strange female who came close to the nest where his wife was sitting and made him proposals of love—and then on the very same day was seen to meet this new female on the other side of the lake and succumb to her charms without more ado."

Dr. Lorenz is disposed to find "human weaknesses" in nearly all animals. For instance, he says that his dog Bully was an accomplished "liar." Bully would always run out to meet him with exuberant affection at the front gate, but he would also run out there to bark savagely at strangers. In old age Bully's eyesight grew dim, and one day when the wind was blowing the scent in the opposite direction, recognition failed and Bully barked fiercely at his master. When he got near enough to perceive his mistake, he stopped short, then rushed past him and across the road, where he pretended to be barking at a neighbor's dog who was not there.

Cats, too, are exceedingly vain of their poise and dignity. If by some chance—perhaps a slippery floor when they are in a hurry—they slide sideways, they will instantly turn to examine some object in the new direction, giving a careful smell to each detail, as though that had always been their intention.

"Women and children first," a precept of chivalry not invariably lived up to by the human race, is an instinct that can be relied on absolutely in dogs. The most ferocious dog terrorizing a neighbourhood of canine males will never touch a female or a young puppy. If a neurotic spinster bitch should attack him, he is completely nonplussed. His pride prevents him from running away, but he cannot bring himself to give battle. So he just stands about, shifting from foot to foot like a bewildered schoolboy.

Magnanimity to the vanquished, another high standard of conduct adhered to on occasion by civilized man, is a law of Nature among wolves. When there is a fight between two timber wolves and the weaker is beaten, he stands rigidly still, turning his head in such a manner as to expose his throat deliberately, the primary point of attack. It is a gesture of surrender, a plea for mercy, and it renders the victor quite unable to attack. The victim, so long as he holds that position, is safe. This same instinct of magnanimity—as it may be called—is to be found among many kinds of dogs.

Pleasure in owning property is an instinct extending far down into the animal kingdom. It is this pleasure, most often, that birds are proclaiming when they sit on a high treetop and sing. They are shouting: "This is my territory. Trespassers keep out!" Other animals deposit a proprietary scent along the borders of their private estates. The mongoose has a special gland that exudes a tiny spray which he uses for this purpose; if you wipe his markers away with a wet cloth, he will promptly come back and renew them.

A zoo director told me of his attempt to mate a pair of leopards. He kept them in two cages separated by bars until they had fallen quite madly in love. When, however, he admitted the female into the male's cage, the male's property sense overrode both love and lust. He snarled and struck her dead with one blow of his paw.

And so it seems that it was Nature, not man, who invented the delight of owning a little piece of this planet we live on.

From: *Animals Can Be Almost Human,* by Max Eastman.

Friendly Dolphins

There have always been some creatures, however, with whom man throughout the ages seems to have enjoyed an especially close relationship. One of these is the dolphin. From time immemorial tales have been told, for example, of the ways in which dolphins reputedly come to the help of fishermen.

One of the earliest accounts is that by Pliny the Elder who about the year 70 A.D. was procurator of Gallia Narbonnesis (Provence). In the region of Nimes there was a marsh called Latera where dolphins caught fish in partnership with human fishermen. At a regular season shoals of mullet rushed out of the narrow mouth of the marsh into the sea. The turning of the tide made it impossible for nets to be spread across the channel and the eddies helped to hasten the escape of shoals of mullet. The dolphins barred the passage to the sea and they drove the scared mullets into the shallows. Not only did the fish provide a meal for the dolphins, they received pay from the fishermen in the form of a bread mash dipped in wine.

Pliny also describes how a dolphin used, every day, to take a boy to school on his back across the Bay of Naples. There are plenty of recent stories almost as astonishing. It's only a few years ago that a young bottle-nosed dolphin became a great tourist attraction at Opononi beach in New Zealand, mingling freely with the bathers, playing ball with them, and allowing children to take rides on her back. Another famous large dolphin which frequented New Zealand waters was Pelorus Jack: for over thirty years up to 1913 it is said to have met regularly ships in the Pelorus Channel, in the South Island of New Zealand, and "escorted" them up the narrow Marlborough Sound. Pelorus Jack became so well-known, and was regarded with such affection, that he was officially protected from molestation by a special Order in Council. There are also many stories of dolphins coming to the rescue of people in difficulties in the water by nudging them along, or pushing them up to the surface if they have gone under. This form of action would come naturally to the dolphin because the mother dolphin always prods her newborn baby up to the surface to help it draw its first breath.

The instinctive social behavior of dolphins, in fact, seems to predispose them to friendly relations with man. It's almost as if they regarded us as a species of terrestrial dolphin. And of course we all know how quickly dolphins in captivity learn all kinds of tricks to entertain us—and enrich their owners. The dolphins thoroughly enjoy the performances, because they have a highly developed propensity for play—as many sailors have noted when, in their natural state, they take the greatest delight in racing alongside the ship or riding in its bow-wave.

Nowadays the friendliness of the dolphins is being exploited by man—and not always in a pleasant way. They are being trained, for example, to find nuclear weapons lost at sea, and then to attach retrieval gear to them; to stick limpet mines, in time of war, to the hulls of enemy ships; and to carry explosives so that they act as living torpedoes—at the cost, of course, of their own lives. It seems a mean way, to put it mildly, of responding to their friendliness.

It is the intelligence, however, of dolphins that is truly amazing. First, they have a brain capacity actually larger than man's. Second, now that we are beginning to record and analyze their communications with each other, scientists have discovered that the dolphin appears to be able to remember a sequence of notes which can last up to *half an hour* long and then repeat the sequence exactly. This is equivalent to an actor learning a major role for the theatre!

What they are actually saying to each other we do not yet know but serious attempts are being made at marine institutes to create a Dolphin/English and/or English/Dolphin dictionary—the first ever interspecies true communication. It's an exciting possibility!

Often, in fact, one can't help feeling that the dolphins show up in a much better light than their human captors. Being sociable animals, they cannot stand solitude. Left alone, a captive dolphin finds it dull and stops feeding. He actually pines away to the point where he may die, but bring him a companion to stimulate his interest, give him something exciting to do and he comes to life again. Not only do dolphins know passionate love founded on sex, but dolphins also show friendship, a pure spiritual affection whose pleasures lie only in the joys of companionship. An illustration of this love was shown once by two friendly males who lived in perfect harmony in a pool which they shared with a number of females; one of them was taken from the pool to star in a series of shows elsewhere. Three weeks later he was brought

back. The two friends' display of joy was truly explosive. For days they remained inseparable and ignored their consorts completely.

In another experiment, in 1962, Dr. John Dreher, a Californian acoustics expert, was conducting research on the Pacific grey whale. This involved a complex device consisting of a series of aluminum poles, cables and microphones. The underwater microphones picked up the signals of groups of dolphins. The signals were evenly spaced. At a distance of 400 yards the dolphins stopped and appeared to gather together, all emissions ceasing. Then one dolphin broke away and came to inspect the obstacles by sonar. He then returned to the waiting group and the microphones transmitted what seemed to be a general discussion. Then the dolphins decided that the strange poles were harmless, for the group resumed its course and quietly passed through.

Although it is true that we must not think of animals as if they were humans, or apply to them moral judgments which belong only to human situations, there's no reason why we shouldn't sometimes find in them patterns of behavior which we would also like to see in humans. In this sense even the killer-whale can display admirable characteristics. Robert Stenuit in *The Dolphin: Cousin to Man* gives such an account.

During a recent Antarctic expedition, a Norwegian whaling fleet received a radio call for help from a deep-sea fishing fleet. Several thousand killer whales *(Orcinus orca)* had arrived in the fishing area and were thoroughly decimating the fish. The whalers sent out three boats, each equipped with a harpoon gun. One fired a single shot which wounded or killed a whale. Within half an hour all the whales had completely disappeared from the surface of the sea around the gunboats but remained just as active around the fishing boats. The wounded whale or other whales who had witnessed the incident had immediately spread the alert, described the danger and even specified the dangerous zones. This story illustrates their organization and solidarity on a level with that of humans. But more important, it shows their ability to transmit messages to deal with an unexpected peril from a modern object totally alien to their own environment.

Pigeons and People

Tony Soper

While the relationship between man and an intelligent animal like a dolphin seems feasible, the harnessing of pigeon power is somewhat more bizarre!

Those rather scruffy, badly-fed birds that strut about foraging for scraps amongst the dirt and grime of our streets came originally from a surprisingly different habitat. The natural homes of our street pigeons are the wild cliffs of the sea coast.

Nowadays you have to go to Scotland to be sure of a chance of seeing rock doves, but once they were common around the coasts of Great Britain. They nested in the rock crevices of the cliffs and foraged along the pastures of the clifftops for food. Now the street pigeon descendants of the rock dove forage in city parks and streets for food, and they nest in the cracks and crannies of city buildings. They have learnt to depend to a certain extent on people for their food and shelter.

Pigeons have lived in close companionship with people for thousands of years. It's probable that they were first domesticated in Neolithic times. The ancient Romans built elaborate rooftop towns specially for pigeons, but although they were fond of them they also had a very practical purpose in mind: they liked to eat them. Young pigeons are rather ugly, but at four weeks old they are fat and delicious.

So the Romans built stone towers with special nesting holes around the inside walls and they employed men to look after the pigeons and farm them for the table. In fact, one of the guardian's jobs was to half-chew bread so that it could be fed to the pigeon-squabs in the most convenient form. In those days farmers had to kill off most of their sheep and cattle stock in the winter because they hadn't enough food to keep them. They just managed to feed enough to keep a breeding nucleus.

The habit of keeping pigeons for food spread through Europe. In France, for instance, there are thousands of dovecotes still standing. Usually they stand on a hilltop, where they command a good sweep of country suitable for the pigeons to forage over. In Great Britain you'll find them near manor houses or monasteries, because the law used to be very strict, allowing only certain important people the privilege of keeping pigeons. A dovecote might have as many as 500 nest holes, and might produce as many as 200 fat squabs every week.

But today the dovecotes are mostly neglected, except for the odd few which are preserved as museum pieces. The reason for this is quite simple. In the eighteenth century came the discovery of root crops, and farmers were able to feed their sheep and cattle right through the winter, using turnips. Fresh meat thus became available at any time of the year, and nobody needed to bother with pigeons any more.

But of course pigeons have other qualities besides being good to eat. They are graceful on the ground and attractive in flight. In fact there was once a Chinese man who loved pigeons so much that he arranged for them to provide him with music "on the air." He attached extremely light whistles to their tail feathers. The whistles consisted of three or four reed tubes of different lengths rather like a pan-pipe. When a flock of pigeons circled in the air, the wind whistled through the pipes and produced an open-air concert. The Chinese liked to say that these sounds were supposed to keep the pigeon flocks together and to protect them from hawks, but it seems much more likely that the pipes were put on simply because the pigeons' owners liked the sound.

The Chinese also pioneered the use of pigeons to carry messages. A poet called Chang Kiu-ling corresponded with his relatives by means of a flock of homing pigeons which he trained in large numbers and called his "flying slaves." The messages were attached to the feet of the birds, but he says they were difficult to train and that it was perhaps two or three years before they could be employed for long-distance flights.

Merchants and businessmen soon began to realize the usefulness of pigeons and they trained them to carry news of prices or the arrival of cargoes. Right up to the time of the introduction of the telephone, homing pigeons were used to send quotations of money exchange rates from the banks in Canton to those in Hong Kong.

Pigeons were also taken to sea by the early navigators. If they became lost, they would set free a bird in the hope that it would show them the way to land.

In Egypt, pigeons were once used in a most unusual way. The Caliph Aziz was a distinguished scientist who happened to be very fond of cherries. And his friend the Yakub Ben-Kilis dispatched 600 pigeons to him, each of them carrying a small silk bag attached to each leg containing a cherry. That was the first known example of parcel post!

Pigeons have also been used extensively in times of war. Probably the most famous occasion was during the siege of Paris, when, in 1870, the only news that came in from the outside world was by way of homing pigeons. Fortunately, before the siege began, people had had the sense to set up some pigeon lofts right inside the city walls.

While the armies of Bismarck completely surrounded Paris they couldn't stop the pigeons flying high overhead and into the beleaguered city. It is said that 115,000 official dispatches and about a million private messages reached Paris by pigeon mail. The dispatches were microfilmed, so that twelve to sixteen large folio pages of news could be contained on an area of about a hundred square centimetres. The contents of a complete issue of *The Times* could thus be sent by one pigeon.

The films were rolled and placed in a quill which was sealed and fastened to a tail-feather of the pigeon by means of a fine wire. The people wanting to read the incoming microfilms would project them, very much enlarged, on to a screen.

Although the pigeons would fly direct to their home lofts in Paris, they wouldn't carry messages in the reverse direction, away from their homes. The only way out of Paris during the siege was by balloon, and there was an almost regular service out of the city and over the heads of the besieging Prussians. In company with the two or three balloonists and their equipment, all escaping balloons carried a full wicker basket of pigeons, ready to be used again as message-carriers when they landed safely.

It's not easy to navigate a balloon and some of the pigeons certainly landed in the wrong place and found themselves in the hands of the enemy. One of the balloons even tried to cross the English Channel, but landed in the sea, not far from the coast. The unfortunate balloonist drowned, but not before he had opened the pigeon cage and released the birds to fly to freedom.

During the First World War pigeons were used a great deal to carry messages back from the front. The army had its own Pigeon Service, and London buses were converted into mobile homing lofts and situated at a safe distance behind the lines. Motorcyclists took the birds to the trenches. When the time came, officers could write messages on a special pigeon-message pad, and attach them to a capsule on the bird's leg.

There were many cases where birds did noble work under terrible conditions. One such was on October 21, 1918, when a homing pigeon was released with an important message at Grand Pré at 2:35 p.m. during intense machine-gun and artillery fire. This bird safely delivered its message to the loft at Ramport, a distance of 24.84 miles away, in twenty-five minutes. One of its legs had been shot off, and its breast was injured.

A special medal, the Dickin Medal, sometimes called the Animal VC, was struck to reward exploits like these. Pigeons were still being used extensively in the Second World War, and in fact thirty-two were awarded the Dickin Medal. A Blue Cock called Billy won his medal in 1942, when he was released by the crew of a force-landed bomber in the Netherlands. He flew through a gale-driven storm of snow to deliver his message safely the next day. Billy was in a state of complete collapse, but the message had got through.

Even in the beginning of the twentieth century there was a genuine pigeon-mail between New Zealand and inhospitable Great Barrier Island, seventy miles away. A man called Fricker hit upon the idea of a permanent daily pigeon-mail with Auckland, because the mail-steamer called only once a week. The letters had to be written on a special form, and the postage was 2½p (5c) one way, 5p (10c) the other. Later the New Zealand government replaced this service with an official one.

Surprising records of speed and endurance have been achieved by pigeons. The world distance record is 2,039 miles, won in 1937 by a pigeon which raced from Manitoba to Texas in 43 days. In good weather young birds will fly about three hundred miles in from seven to nine hours, and flights of more than six hundred miles have been made in one day by older birds. Some pigeons can fly five hundred miles without stopping to eat or drink. The distance from Dover to London (76 miles by rail, 70 miles as the pigeon flies) was once covered by a homing pigeon, beating by twenty minutes the express train which ran at sixty miles an hour.

From: *Animal Magic,* ed. by Douglas Thomas.

The Camel—Discontented Ship of the Desert

Arthur Weigall

Not all animals, however, have an easy relationship with man, as this delightful description of the undervalued camel shows.

All camels are discontented. They hate being camels, but they would hate to be anything else, because in their opinion all other living creatures are beneath contempt, especially human beings. The expression upon their faces when they pass you on the road indicates that they regard you as a bad smell.

They nurse a perpetual grievance against mankind and ruminate upon their wrongs until they groan aloud. When you go to them to find out what is the matter they give you no hint of any specific trouble, but merely look at you with sad, reproachful eyes and groan more loudly. In certain cases when their sense of unbearable insult is overwhelming, they try rather halfheartedly to bite you.

The fact that a camel has yellow teeth, a harelip, a hump, corns and halitosis places the poor creature beyond the range of ordinary sympathy. People never put their arms around camels or stroke or kiss them, and yet their sorrowful eyes, fringed with long, languishing lashes, are beautiful, and their whimpering is heartbreaking. But camels do not ask for love or pity. They make no response whatsoever to overtures of that sort. They have no hope, and they make no friends. When they are being ridden they do not attempt to cooperate with their riders, and when they are being used as beasts of burden they try their best to make you feel a cad.

In the days when I was Inspector-General of Antiquities in Egypt, I had a horse, a donkey and a camel: the horse for ordinary riding; the donkey—one of those big white Egyptian donkeys which are almost the size of a mule—for making my daily rounds among the temples and excavations in my charge; and the camel for desert work. The camel's name was Laura, but she neither knew nor cared.

Laura proved to be one of the strongest runners of her kind. I rode her on several long expeditions into the Egyptian desert, and on one of these trips she carried me 200 miles in four days without drinking. But that is nothing remarkable for camels can go ten days without water and can keep up an average pace of eight to ten miles an hour for six or eight hours during the first three or four days. When properly watered, they can run 100 miles a day, though 60 is generally considered a good, hard journey.

Like all camels, Laura was supremely stupid. For instance, she could never be taught that she must remain crouched until her rider was in the saddle, and must not scramble to her feet just at the moment when he was mounting. A camel's saddle is a sagging square of leather covered with a thick sheepskin, supported on a high, padded framework which fits around the hump and is fastened by a girth under the animal's body. You sit on this lofty throne with your legs crossed in front of you on the curve of the camel's neck, which is your footstool. Once you are up, you cannot get off again until you have made certain strange noises with your tongue, the signal for the camel to kneel. Then down go the awkward creature's front legs, and you hang forward over its neck. After that the hind legs double up, jerking you straight again. You can then slide gracefully to the ground—unless the camel suddenly decides to get up, in which case your descent is more spectacular.

Laura always watched me out of the corner of her eye until she caught me at a disadvantage. When I swore at her she only gazed at me sorrowfully and uttered her inconsolable grumbles. A camel, by the way, can do more than look at you out of the corner of an eye. It can turn its head completely round and stare at you full in the face with both eyes. I know of nothing more disconcerting.

There are two kinds of camel. One is the Bactrian, which has two humps and long, shaggy hair; but the kind used in Egypt is the Arabian, which has only one hump and short, sandy-colored or whitish hair. The hump is a store of fatty flesh and skin. Thus the camel's body can cool more rapidly than a man's, for example.

The hump, then, is one of the reasons why the camel can survive great desert heat. Two other factors are important, too: the camel can raise and lower its body temperature and it can tolerate dehydration. Camels show themselves to be desert creatures in additional ways: their flat, padded feet are designed for treading on soft sand or hard rocks, and they can close their nostrils during a sandstorm.

It is not customary to allow a riding-camel to walk, because the motion is rolling and you lurch from side to side in a sickly manner which suggests a reason for calling the camel the "ship of the desert." A quick jog trot is the usual gait; you simply bump up and down in the saddle like a cavalry trooper, the bumps becom-

ing bigger and better as the pace increases to a gallop. Laura used to add to the fun by occasionally jumping over low rocks, but she never fell. In fact, I have never heard of a camel falling.

Laura became a mother when she was about ten. In the spring the male camels attract the attention of the females by making gurgling noises, like water running out of the bath, and inflating their tongues until they hang out of their mouths like pink balloons. Laura could not resist the blandishments of a magnificently disdainful he-camel who hailed from down Suez way and was in the transport business. As she appeared to be all wrought up in her own melancholy fashion, we arranged a rendezvous. Although Laura was not a large animal as camels go, the resulting foal stood three feet high when it was a week old.

I have sometimes heard it said that camels are delicate and difficult to rear, but in my experience camel health is often remarkably good considering the poor food they usually get. The birth of Laura's foal gave her no trouble, and the foal itself was healthy enough. Friends of mine in the army have told me that the camels used in desert warfare do not seem much distressed when they are hit by bullets, and recover rapidly from wounds. I suppose this depends on the camel. Even with good food and care, Laura had colic sometimes, and in winter she would catch cold and mope about with her nose running. As for pain, she nearly wrecked our hospital at Luxor when an abscess in her hump was lanced.

Laura's various expressions of loathing, together with endless groans and complaining, made you think she could not possibly be in good health. I used to watch her teaching her foal to grumble. When she saw me coming she would start bleating and bubbling, putting her head close to her infant's as she did so, in order that the sounds might be imitated.

Heaven knows I treated her and her offspring with kindness, but I never saw the light of love for me in the eyes of either of them. Laura did love her foal for the first few months of its awkward, leggy little life, yet in the end she gave it a nip and the small creature kicked her in the ribs in return. After that they went their separate, dreary ways with their noses in the air and their hearts full of their grievances.

From: *Laura Was My Camel,* by Arthur Weigall.

ANIMAL MARVELS

Adaptation Techniques: Bat versus Moth

Vitus Droscher

Perhaps it is an indirect result of the danger of ex-termination that faces so many species in so many parts of the world, that in recent years research into the be-havior of all the various forms—insects, fishes and rep-tiles as well as birds and mammals—has been con-ducted with ever-increasing intensity, and a mounting sense of excitement. For all the mysteries that have been solved, as many amazing discoveries have been made. This chapter will touch on a few of them.

A great deal more has been learned, for example, about the techniques evolved by creatures of all kinds in adapting themselves to the challenges of their par-ticular environments. Often these are of the most astonishing subtlety and complexity. Take the con-stant warfare that takes place between moths and their predators, the insect-eating bats. No human military

campaign could be more packed with camouflage and evasion, traps and deceptions, stratagem and counter-stratagem.

At first sight the odds seem to be all in favor of the hunting bat, for (in most of the species) he is equipped with an echo-location system which makes those of a modern air force look crude by comparison. In the bat's mid-brain, scientists have discovered, there is an auditory complex that can recognize sonic impulses as separate sounds even if the interval between them is no more than a thousandth of a second. In consequence it can register as separate noises sonic impulses which dart through the air at a distance of only thirteen inches from each other, and can therefore distinguish by the echo alone two objects separated from each other by that distance. What is more, only one thousandth of a second after a bat has sent out its exploratory noise (a click or a high-pitched whistle) its ear is able to receive back the first echo. This means that it can accurately locate a moth or some other insect when it is only a few inches away. With an echo-location system like that it hardly matters that the insect-eating bat's tiny eyes can see practically nothing in a dim light—and in fact, attempts are being made to see if some aspects of the bat's echo-location mechanisms can be incorporated into devices to help blind people get around.

The moth, though, has evolved methods just as sophisticated to cope with the situation. A remarkable book, *The Magic of the Senses*, by the German zoologist Vitus Droscher, best describes the complicated chess-moves in the Bat versus Moth epic.

∽∾∽

By their phenomenal radar sense the bats change the darkness into daylight for themselves. Thus they dominate, almost without competition, a rich living space: the night with its abundance of buzzing and humming insects. One would think they must be really in paradise. But many insects have developed a number of tricks and counter-weapons which make things more difficult for the bats.

An insect's first counter-measure, as it were, is its own body; for while the bats are eating that, it effectively gags them and thus prevents their getting their bearings. But the little brown bat has an answer to that. While chewing and swallowing, it keeps open a sort of gap between its teeth, if at all possible, through which it can "whistle" almost as strongly as ever. But if the prey is too fat, then a substitute transmitter goes into action: the creature emits noises through the nose. Although these are shorter, of a deeper pitch, and only just half as loud, they at least save the bat from colliding blindly with trees or other bats. This works even better with big-eared and barbastel bats. When they want, they can produce as loud and piercing cries through their noses as through their mouths.

It is the horseshoe bats who in all instances emit their sounds through the nose. These bats are so called because they have transformed the outside of their nose into a trumpet the shape of a horseshoe. The organ acts like a parabolic mirror whereby they can focus sound much like a radar device does with electromagnetic waves. The best comparisons with humans to describe this phenomena is that given by Professor F. P. Mohres who writes "Just as to us humans a landscape becomes visible at night in full reflection of a searchlight, so to a bat its environment

hidden in darkness becomes recognisable in the echo of its sound projector." Sound waves instead of light waves convey to these bats a perfect graphic impression of their environment. Furthermore, the horseshoe bats have a totally unique principle of location. They do not send out intermittent ultrasonic bursts but sequences of ultrasounds lasting much longer on a frequency of 85 kilocycles.

So the weapon of gagging has been ingeniously parried by the bats. Consequently various insects defend themselves in another way: they make themselves inaudible. Besides not making any noise themselves, they are wrapped in "acoustic camouflage"; that is, in such a soft and sound-absorbing integument that it hardly reflects even the ultrasound of their enemies. Flies and mosquitoes multiply so boundlessly that losses through bats threaten the survival of the species as little as do the windscreens of cars. But bats might be a serious threat to the survival of night moths, as a species, had these not escaped the danger of complete extinction by developing absolutely noiseless flight—even in the realm of ultrasound.

According to Professor Heinrich Hentel, the Berlin biophysicist, moths have had to avoid turbulence at the edges of their wings to prevent them from giving themselves away. Night moths solved this by fine fringes made of tiny hairs only about two millimeters long which prevent the formation of air eddies that produce the noise. However, for many species of night moths even this way of flying is still not silent enough, so they have developed special ears for listening to the enemies' transmitter. The night moth's ultrasonic ears consist of two sensitive auditory nerve cells that can pick up the bats' ultrasonic cries. In consequence, immediately when a bat comes within a hundred feet of a flying night moth, the highly sensitive nerve of the moth's ear registers the bat's cries, and a series of signals sends a preliminary alert to the brain. The moth at once swerves onto an avoidance course, in the opposite direction to the bat. If the bat is below, the moth will even zoom straight upwards.

Although it flies much more slowly than the bat, this alone often proves a life-saving maneuver, for by that time the bat's sonar has not yet located the moth. Because of the sound-absorbing padding of the moth's acoustic camouflage, the bat can only "pick up" its target within a range of twenty feet.

Some moths are capable of even more sophisticated evasive action. The experiments of Professor Roeder have shown that at the noise of an artificially produced ultrasound some moths looped the loop and others performed aerial acrobatics to make the bats miss them. As the bat flies along, moths several hundred feet from its echo-location source will begin to scatter. If the bat flew as straight a course as the swallow, it would scarcely ever catch a moth. So it has developed a counter-strategy, a "reeling" flight. This appears clumsy, but that is an illusion. All the curious swerves are in fact different geometrical curves broken off abruptly by the bat to deceive its victims about the direction of its flight!

From: *The Magic of the Senses*, by Vitus Droscher.

The Changing Chameleon

Some creatures employ color as a means of outwitting their predators. Many lizards have the ability to change their color and in this respect, the most famous is undoubtedly the chameleon. It is commonly believed that the chameleon simply changes color to match that of its background. This is not so. In fact, the change of color is limited to three colors: green, yellow and brown, or perhaps a mixture of two or all of these. In addition, some species (but not all) can produce quite a large number of black dots. Of course, this is the range of color valuable to the animal in its natural state. No chameleon can change to blue or red or black and white.

Another widespread belief is that the color change in the chameleon is virtually instantaneous: take a chameleon from a green leaf and place it on a brown table and it will at once change from green to brown. This, too, is not so. No one can say precisely how long a chameleon will take to make a color change. So much depends on circumstances, upon light and temperature, and upon such imponderables as anger or fear. The time taken for color response is normally short. It may be only a matter of seconds, but under experimental conditions it usually takes a minute or two and sometimes very much longer.

How is the change accomplished? Chameleons have four distinct layers in the skin. At the base of the skin are pigment layers, known as melanophores, containing black-brown melanin. Above this is a layer which, owing to the presence of gyanin crystals, appears white by reflected light. This layer provides a background against which the colors of the two upper layers can be well displayed. The top layer of all is of yellow pigment. The yellow pigment of this top layer can be spread out to make a common layer or can be concentrated so that the light passes down to the layers beneath. The melanin on the bottom layer can be withdrawn into the cells at the base of the skin and so concealed altogether or it can be passed up to the layers above through hollow branches, or it can be spread over the whole surface of the skin sufficiently densely to conceal all the colors beneath.

This mechanism is brought into play by light and is a reflex action. The chameleon responds to the color tone of its background, so that it becomes pale on a pale background and dark on a dark background. If one part of the skin is more strongly lighted than another, that part will become darker than the other. Thus chameleons—and all other lizards that undergo color changes—become pale at night.

Cold Storage Mice

Many animals also possess the quite astonishing ability to adapt themselves to drastic changes in their environment, often within a comparatively short space of time. The ordinary house mouse is an outstanding example. It had frequently been observed that putting perishable foodstuffs into cold storage rooms does not keep the mice out—and that they could live quite comfortably at temperatures of 10° C., feeding entirely on meat frozen solid. What is more, these mice reproduce throughout the year, although their fertility is reduced and many of the babies die.

Researchers at Glasgow University, however, have found that mice can adapt to even severer conditions.

They reared fourteen generations of mice at a temperature of only 3° C. They found that the first two generations had shorter tails and heavier hearts and stomachs than their parents, the result, presumably, of a higher metabolic rate. They also had thicker fur, especially the females. Then came a great surprise: the succeeding generations showed none of these features, and were almost exactly like the ordinary mice reared in the warmth. In other words, their physiological adaptation to cold was by then so complete that they no longer needed any kind of anatomic specialization.

The Incubator Bird

Various species of birds have very frequently been compared to human builders, but the most remarkable of all the bird builders is perhaps the incubator bird. It was given this name because it builds itself an incubator, scratching together some leaves and grasses, and hatches the eggs in the warmth of their decomposition.

The egg chamber of the mound must always be maintained at a constant temperature of 33°C. This involves the bird every day for a period of six months whereby air vents have to be dug or closed, a covering of sand as heat insulation has to be removed or put on, thickened or thinned out. The bird's sensitivity is so highly developed that it would gauge the temperature inside the mound to within a tenth of one degree.

Playing Possum

A number of creatures make use of a kind of trance or temporary paralysis in order to evade their enemies.

Under certain circumstances playing possum has a high survival value. The most accomplished feigner of death is the American opossum. In normal sleep the opussum keeps its mouth and eyes closed and feet out of sight, but when attacked it collapses with eyes open, lying on its side. Tests have shown that the animal is still wide awake because it responds to loud noises by twittering its ears, and there is no difference in oxygen consumption, body temperature, or blood chemistry; an EEG recording shows that the brain waves are identical to an alert animal. So this species has developed a state of automatic paralysis to inhibit attack and perhaps give it a chance to escape.

Sleep in Animals

Dr. Lyall Watson in *Supernature* gives some fascinating information about sleep in animals. Cold-blooded species become inactive during the cool of the night, when their body temperature falls as fast as the air temperature, while birds and mammals have developed an independence from this by controlling their internal temperature so that they may be active in the dark. Most birds sleep with their eyes closed and their head tucked underneath a wing. Dolphins appear to sleep with first one eye open and then the other. Even animals like elephants and giraffes that are traditionally supposed never to sleep do in fact slumber, often lying flat on the ground as they doze.

The Senses of Animals

Many of the mysteries and marvels of animal creation can be explained by the extraordinary development of whichever of the senses the individual creature needs for survival. The most obvious example, perhaps, is the dog's sense of smell. There have been so many instances of dogs tracking down criminals or hidden caches of drugs when there was no apparent scent-track to follow that at one time many people thought that the dog must possess some other sense, so far unknown to us. One dog-breeder, for instance, pointed out that his dog could find him even when he was wearing rubber boots—which must surely be scent-proof. Professor Neuhaus of Germany, however, was able to prove that even rubber can let through a scent trace sufficiently strong for a dog to detect, though it would be utterly impossible for a human nose to do so. In fact it has been claimed that a dog's sense of smell is a million times better than ours.

The bee's eye and a bat's hearing also provide two very good examples of the specialization of the senses. A bee's eye divides the sky into a screen of squares. An eye would have 15,000 eye facets and each would observe its own sector of the picture. At each moment of time the sun is seen only by a single lens which provides the insect with an almost perfect device for measuring the angle of its direction in relation to the sun and take its course in relation to the sun's position.

The bee's eye not only creates a navigational instrument but also a device for measuring its flying speed. Each lens registers a change from light to darkness, and the next lens or the one after it registers the same change a short time afterwards. From the difference in time the bee's brain works out the flight speed over the ground.

Animals can respond with the whole of their organisms to their environment—and not only the terrestrial one. Many animals are sensitive to water. In times of drought elephants are called upon to divine hidden water sources by using their tusks and feet. Two suggested solutions have been put forward to explain this sense. First, that they can smell the water percolating through the soil and second, they have an elementary understanding of geology in that they always dig in the lowest point of the outside curve of a dry river bed where water is most likely to collect. Or perhaps some other sense is involved whereby their bodies can "fine tune" in to possible water sources. Certainly it is logical that animals, whose bodies are composed of such a high proportion of water, should be sensitive to this essential element.

Learning in Animals

To what extent, in adapting themselves to changes in their environment, do animals "learn," in the sense that we can apply the term to human beings? Much more, probably, than it was thought at one time. No doubt we all know that a chimpanzee can work out for itself, by a process of trial and error, how to get at a bunch of bananas placed out of its reach—sometimes, in fact, by quite sophisticated methods, such as piling boxes on top of each other or by joining slotted sticks together.

Dr. Lyall Watson gives an account in *Supernature* of how young squirrels encountering a hard-shelled nut for the first time will make indiscriminate scraping patterns on it with their teeth until the nut yields and breaks open. As they gain more experience they learn how to apply the minimum of effort for the maximum return by following fibres in the shell and not working against the grain.

Crisis in a Beehive

The entomologist, Professor R. Darchen, conducted an experiment to see how quickly bees could take counter-measures in the face of a sudden and unexpected disaster. He took a nesting-box, in which the bees had just completed a new honeycomb wall, and tipped it on its side. The older walls were sufficiently set to stay in place, but the new one, which was of course still soft, began to bend, threatening to stop up the entrance to the breeding-chambers of the comb below it—and so eventually kill the brood growing up in it. But the bees went into action immediately to prevent the wall from bending any further. Hundreds of them rushed to the spot where the bulge was at its greatest, and with their legs stretched out braced themselves against the pressure of what to them must have been quite a massive weight. They had in effect used their bodies to form props, similar to those used by human builders in shoring up a weakened wall. While this was happening, hundreds of other bees were frantically at work making pillars out of wax, to substitute for the living props.

Another entomologist, Professor R. Chauvin, commented that Darchen's papers are of particular interest because they have shown us that some traditional views about animals being subject only to their instinct do not agree with the facts. The idea that bees work like machines regulated in advance once and for all is thus shown to be completely wrong. In reality it is all much more complicated. We are dealing with a social group, which recognizes difficulties and can solve the problems facing it. So here is no mechanical process dependent of the instinct, but an activity of a higher order.

Language and Communication

It is true that Professor Chauvin goes on to warn us that the word "intelligence" is not really applicable, "because for that phenomenon a number of other things are needed as well"—and intelligence can be regarded, in many respects, as a technique of adaptation belonging exclusively to man. But undoubtedly "activity of a higher order" than mere blind instinct is frequently at work among animals to a far greater degree than was realized in the past.

This realization has bearings on the ways animals communicate with each other, and with man too when they are in contact with him. "Language" among animals is, again, a much more complex matter than it was at one time considered. It is not only scientists conducting special research into the subject who have come to understand this, but quite often those who have lived close to an animal in their own homes.

Busy Bees

Dr. Harold Esch of Munich University discovered that worker bees communicate with each other by making various rattling noises from their wings as to the source of their feeding place. The richer the source of nectar and pollen the faster the bees rattle.

Talking Ants

The American zoologist, Professor Edward O. Wilson of Harvard University, has found indications of a scent language in insects. He has already deciphered words of an apparent dictionary of odorous substances and even suspects that ants may have a kind of syntax. With ants alone at least ten scent words have been deciphered. Experiments with the American fire ant have shown that the fire ant pushes its sting out of its abdomen and presses on the scent gland which makes a fine trickle of liquid run along the sting like ink on a fountain pen. In this way the fire ant "writes" a scent on the ground with its sting. The scent rising into the air from these ants is capable of attracting other ants at a maximum distance of 4 or 5 inches. However, each ant species has its own secret scent which cannot be traced by members of a foreign species.

Noisy Fish

Fishes communicate with each other too. It used to be thought that they lived in a silent world—but this is far from being the case.

In 1954 Dr. Hans Hass, the Austrian underwater explorer, conducted experiments with an underwater microphone which he let down among the fish at a coral reef in the Caribbean. The loudspeaker on the deck of his sailing ship produced a multitude of noises, ranging from buzzing, whistling, and knocking to a variety of rhythms. Human ears do not hear these sounds because we are adapted to hearing in the air, while in water the acoustic conditions are different. The volubility differs from species to species. Ocean fish make more noise than fresh fish, while those from tropical waters have a richer vocabulary than those in northern waters. Fishes' voices also get deeper with increasing age.

Generally fish make their noises by drumming on their swim-bladders with their muscles and ligaments. Very few use their mouths for making noises. Trigger fish, moon fish and soldier fish produce creaking, squealing, and screeching noises by rubbing together incisor teeth. But there are other methods of communication besides sound. Following are some of the more surprising.

Horse Sense?

Henry Blake

Henry Blake, student of horse behavior, found that a horse employs quite a wide vocal range in communicating both with his own kind and with humans.

Before you can even start to interpret a message made by sound . . . you have to know the sex of the animal. We also discovered that it is important to take into account the age of the horse, because obviously the range of tones and notes used by a foal is completely different from the notes and range he will use as a stallion four or five years later.

On the other hand the stallion, mare, gelding, foal and yearling will all have the same number of tones and notes, and they will be made in eleven different ways. Nine of the eleven different tones of voice are made by exhaling, that is to say they are made by breathing out. First there is a snort, which is made by using the nostrils alone as a sound box, and at times the imperative is expressed by crackling the nostrils

at the same time. The stronger note with the crackling of the nostrils is used to draw attention as a signal of alarm, as a sign the horse is excited or to denote strong emotions. The whicker is also made by using the nostrils as a sound box, but this is a much more caressing note and can vary from a gentle blowing through the nostrils to quite a strong sound, used usually as a greeting or to show affection of some sort. Then there is the whinny, which is a much higher-pitched enquiring sound, and the neigh, which is stronger again than the whinny. In these two the voice box is used. In addition we know the squeal of the mare, and the bell of the stallion, each of which can have a distinctive sexual tone to them, or may sound aggressive or be used as a warning. These both come from the upper nasal regions of the voice box and are used in sex play, in anger or to display temper. The stallion has a whistle which he uses to call the mare; and all horses have a scream of fear, pain or anger which comes as a gust of terror from the lungs. These are all exhaling sounds. The breathing-in sounds consist of a snuffle, which corresponds to the gentle blowing out, and a sniff, which corresponds to the snort. Each of these notes has a definite meaning for another horse.

The stallion has the greatest vocal range, and some of his notes are frightening, while others will be very beautiful to hear. But he has a somewhat limited range of messages to deliver with his voice, simply because in his natural state he is concerned only with three things: sex, danger and food. So his messages are confined to these three subjects.

A mare on the other hand, while she has her sexual sounds and her sounds for food and danger, also has a range of sounds for the care and protection of her foal, and probably her yearlings as well. She has to call her young to her for food, she has to call them in case of threatened danger and she has to teach them discipline, so her range of messages will be far greater than that of the stallion. A gelding, which of course does not exist in the wild, has a vocal range which may vary from that of a stallion, if it has been cut very late, to that of a mare if it is over-protective to the person who looks after it. A foal, equally, will have its own messages and vocal range concerning food and fear; it will have no sexual messages but it will have a range of sounds asking for protection and reassurance, and these will change as he gets older. He will retain some of his foal phrases as a yearling and even as a two-year-old. Then, when he starts feeling a man in his two-year-old summer, and certainly as a three-year-old, unless he has already been cut, his messages and voice will change to that of a stallion; or a filly will develop the language of a mare.

Contact with man, too, increases a horse's vocal range. This makes for complications, since it is almost impossible to differentiate between the messages that are natural to the horse and those that result from contact with man. If feeding is late, for example, a horse that has been in contact with man will whinny or bang his manger, or make some other sign to remind you it is time to feed him. This action is completely unknown to the horse in the wilds, since his food is always there, and he does not have to draw man's attention to the fact that he is hungry. We note from observation that when a horse discovers the messages that he is trying to convey are understood, either by another horse or by man, he will use it again; that is, he extends his own vocabulary. The most extensive vocabulary we have ever come across was that of Cork Beg whom we owned for twenty years, and we observed that other horses who had had little or no contact with man learned phrases from him and thus extended their vocabularies.

Henry Blake has worked out a dictionary of horse language. Here are just a few examples of its contents:

Welcome. This is used to generalise all calls and signs of greeting used between horses, the most common of which is the whicker of welcome. The strength of the call and the vigour of the movement indicate the degree of imperative. The context, and the carriage of the head and tail, indicate the purpose of the welcome.

Who are you? is used by two strange horses on meeting. It is an extension of the "welcome" phrase and is said by sniffing or more usually blowing at each other. The attitude of the two horses towards each other is indicated by the harshness or the gentleness of the blowing and the carriage of the head and tail. This procedure leads to the sub-messages, (1) *I am a friend,* said by continuation of the gentle blowing and other friendly movements, or (2) *go to hell,* a snap or nip by one or other horse, a stamp on the ground with a front foot, a threat to kick, or a squeal.

Come here. This starts as a whicker of welcome rising in the imperative, which may also be shown by shaking the head back and forwards if there is no response. The message may be changed to *if you do not come here, there will be trouble,* which is shown by a threatening movement and will draw the response, *all right I am coming,* usually said by a low whicker.

I am king is the bugle note of a stallion, which is either a challenge, or a call to a group of mares. This will be repeated again as he goes towards the mares.

I am only small. He says this by holding his head and neck out straight, sometimes holding the nose up slightly and moving the mouth as if sucking. When he does this it is most unlikely that another horse will hurt him.

Gangway! This is said by a boss horse by pushing through a herd and laying about the others with his head. On the other hand the sub-message *Excuse me,* is used by an inferior horse trying to pass a boss horse.

Scratch here is shown by rubbing where the itch is. If he is with another horse, he will scratch the other horse with his teeth to show where he wants to be scratched.

From: *Talking With Horses,* by Henry Blake.

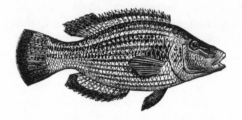

Luminosity

In 1950 the American deep-sea explorer, William Beebe, descended into the depths of the ocean in a specially designed diving bell, known as a bathysphere. His account at various depths revealed a host of animal illuminations resembling stars on a clear evening. Between 2,050 and 2,150 feet he saw relatively few illuminated organisms, but later at 2,200 feet the lights were bewildering in their brilliance.

He watched one gorgeous light as big as a ten-cent piece coming steadily towards him, until, without the slightest warning, it seemed to explode, so that he jerked his head backwards away from the window. What happened was that the organism had struck against the outer surface of the glass and was stimulated to a hundred brilliant points instead of one. Instead of all these vanishing as does correspondingly excited phosphorescence at the surface, every light persisted strongly, as the creature writhed and twisted to the left, still glowing, and vanished.

These brilliant underwater fireworks aren't, of course, merely for entertainment. They're evidence of another whole range of adaptive devices in the struggle for survival in the special conditions of the ocean's depths. William Beebe noticed that when one of these luminous deep-sea fish came very close to the window of the bathysphere he could see the whole of its outline and shape, by the glow cast by its own light. But when it was a few yards away, he could see only a series of light-spots—what he calls a "constellation." He came to the conclusion, though, that this arrangement of lights differed from species to species—and what was meaningful among them was not the glow itself, but the pattern made by the separate points of light. The best human comparison is with navigational lights. At night the captain of a ship at sea knows what kind of craft it is that passes by unseen in the darkness by the kind of lights it carries, and the way they are placed. In much the same manner, these deep-sea fishes, every species of which has its own characteristic arrangement of "lamps," can pick out its own kind, and, more important, distinguish between friend and predator.

The fishes' navigational code is much more complicated than the human one. One species of sea-dragon, for example, to which Beebe gave the Latin name of *Bathysphaera intacta* (the untouchable bathysphere fish) has a row of twenty-four pale blue "portholes" on either side, so that it looks for all the world like a miniature lighted liner. In addition, it has tentacles hanging from its "bow" and "stern." At the end of the tentacles there are lights, trailing in the water. These lights are, in effect, baits—and any creature attracted by them is immediately seized upon by the sea-dragon, whose teeth, incidentally, are also illuminated, from the inside.

Another of these luminous fish described—and named—by Beebe is the *Bathysidus pentagrammus*, or Five-lined Constellation Fish. This is what he wrote about it:

"Along the sides of the body were five unbelievably beautiful lines of light, one equatorial, with two curved ones above and two below. Each line was composed of a series of large, pale yellow lights, and every one of these was surrounded by a semi-circle of very small, but intensely purple photophores . . . . In my memory it will live throughout the rest of my life as one of the loveliest things I have ever seen."

The number of different light displays is tremendous. In the family of the already well-known lantern fish alone, for instance, there are at least 150 displays, according to the different species. In addition, the light displays often differ between males and females—as with the plumage of birds.

What this all means is that the vision of these fish living in perpetual darkness of the ocean depths is restricted to picking out and interpreting a whole range of individual light patterns—patterns as abstract as those of any modernistic painter.

Here are a few examples of the luminous properties of different fish and insects:

WRECKERS

Just as in the past the "wreckers" of England's Cornish coast used to display false navigational lights in order to lure unsuspecting ships on to the rocks (so that they could then plunder the wreckage), so the various species of deep-sea angler-fish carries in front of it a kind of fishing-rod, with a small luminous balloon dangling from the end, which looks just like a luminous marine worm. When the unsuspecting prey darts at the "bait," the angler-fish's razor-sharp teeth are ready to snap it up.

BEARDED FISH

Instead of "fishing rods," the deep-sea barbel fish make use of their long "beards," which have a luminous organ hanging at the end of them. Nerves in the "hair" of these "beards" (which are sometimes nearly a third as long as the body of the fish) signal the approach of the prey. William Beebe, who carried out experiments in an aquarium with one of these fish, reported:

"Even the minutest movement of the water near his beard produced the utmost excitement in the fish. He became savage and snapped, trying all the time to reach the source of the disturbance and to bite."

There's another deep-sea fish, belonging to the viper-fish family, which has about 350 bright dots of light inside its mouth. When it opens its mouth, small fishes and crustaceans swim inside, dazzled by the display.

As for that sudden explosion of lights when some marine creature banged into the window of William Beebe's bathysphere, he found out later that it was a luminous deep-sea shrimp, which releases a shower of flares and sparks in order to dazzle and confuse its enemies. On one occasion Beebe put his luminous watch to the window of the bathysphere when an inquisitive lantern fish was looking in. In its fright the creature reacted in much the same way, by emitting a series of flashes.

And that's only a very small selection of the ways in which deep-sea creatures make use of this device of luminosity.

LUMINOUS BEETLES
Many of the insects which operate at night employ similar methods. The obvious example is the glow-worm—but the luminous beetles of southeast Asia also produce an amazing light show. During the day these insects stay in the jungle but at night the males fly to the mangrove trees along the river banks. There may be thousands of them sitting in these trees and without exception they will all blink and flash in exactly the same rhythm (twice a second) producing a fantastic light show in the night.

Though we do not know the purpose of this display we know that only the males produce this amazing light show.

Migration of Birds

Of all animal marvels the greatest, perhaps, is that of migration. Some amazing examples of migration include the bronze cuckoo from New Zealand. Dr. E. Thomas Gilliard reports that as soon as the female has laid her eggs in the nests of the other small birds, she flies to her winter quarters. Her young would follow her there a month later on their own. To do so they must fly without guidance 1,250 miles to Australia where they pause briefly for rest and food, and then across an enormous expanse of ocean without islands to the Bismarck Archipelago, covering a distance totalling 4,000 miles to rejoin their species.

An even more amazing feat is that performed by the albatross. An experiment was carried out by scientists for the United States Navy when a breeding colony on one of the Midway Islands interfered with the flight pattern of an air base. Zoologists decided to transfer the birds to a place from where they were sure not to return to Midway. As a test, eighteen albatrosses were taken by plane over a distance of 3,000 miles in varying directions. The experiment was a total failure: the albatrosses returned within a short time, the fastest after only ten days.

The Celestial Navigation of Birds

Bird migration still holds many mysteries, but recent research has uncovered some fascinating facts about it. Homing pigeons, a few seconds after being released, appear to know where they are. So far there is no satisfactory answer to how these birds carry out such position finding. But some striking facts have emerged as a result of a series of experiments carried out by Dr. Kramer. Dr. Hans Lohal of the Radolfzell Ornithological Station gives a fascinating account of some of these.

Aviaries were put up at Wilhemshaven in which young pigeons with no experience of flight were held captive. They could not see any more of the world than the view from the cage allowed. Then Dr. Kramer took all the pigeons 95 miles to Osnabruck for their first flight. The novices still managed to find their way home. So they clearly had absorbed their impression of home exclusively from their cage.

In a second experiment Dr. Kramer limited still further the view of other pigeons. He surrounded the aviary with a high fence which allowed the pigeons no view of the surrounding terrain or horizon and left only the sky uncovered. These pigeons were also taken to Osnabruck and none found their way home. On release they were completely disorientated. If pigeons are raised, then, so that they cannot see their surroundings or horizon, they cannot develop the ability to find their way home. It could be that the birds need the horizon as a reference line in the same way as a sailor does to establish the height of the sun without a sextant.

The German zoologist Dr. Franz Souer carried out a series of planetarium experiments which have become world famous. The idea came to Dr. Souer when he noticed that, when the sky was clouded, blackcaps and other European warblers would orient themselves for migration only if a few bright stars were visible. That provided him with a clue that night migrants must navigate by the stars. Experiments began by showing the blackcaps at the autumn migration season an artificial starlit sky which looked exactly like the real sky over Bremen (the birds' home). They were completely deceived by it, and fluttered without hesitation towards their migratory home of Turkey to the south-east. That, by itself, proved nothing at all as the birds might not have taken their bearings from artificial stars but from something quite different, such as the earth's magnetic field.

To meet these objections Dr. Souer turned the pro-jected image of the sky in all directions, but the birds always took off in the direction which should have been southeast by the position of the stars. These experiments made it clear that blackcaps instinctively recognized individual constellations, even when there were only isolated stars visible through gaps in a cloudy sky. But if the whole sky is overcast, they flutter helplessly for some time before deciding that they cannot hope to find their bearing. Furthermore, Dr. Souer's experiment showed that blackcaps *inherit* their knowledge of celestial geography—an act of nature which itself is fantastic.

Robins migrate in autumn too. But as soon as the night sky is overcast and the robin can no longer recognize any signpost stars, it does not, like the blackcap, need to land and wait for clear weather again. It switches over to steering by magnetic compass and flies on unerringly over rivers, lakes and mountains.

The existence of a magnetic sense has been shown by tests on wild mice who have been trained to follow particular compass directions when optical, acoustic and chemical landmarks were excluded. Other tests have revealed that snails can find their way home without being able to see, hear or smell the direction of home. A magnetic sense has been proved for crickets, locusts, wasps and flies as well. So, senses which sound uncanny are more widespread in the animal world than once believed.

Migration of Insects

Birds aren't the only creatures which migrate. During recent years zoologists have started investigating certain phenomena in the insect world which have surprising analogies to bird migration. Butterflies fly distances of 2,000 miles, moths cross oceans in vast swarms and ladybirds (ladybugs) and dragonflies cross whole continents. One of the most amazing migrations is achieved by the copper-colored North American monarch butterfly. It spends the winter in California, Mexico, Louisiana and Florida. With the beginning of spring the butterflies wake from their numbed state. Gradually the swarms move in a northerly direction and two months later they reach Canada and Hudson Bay. From September onwards the butterflies return again to the warm south. This species can cover a distance of at least 2,500 miles in their lifetime. This exacting journey is not prompted by a shortage of food or a change for the worse in living conditions. As with migrating birds there is a strange urge to be on the move which is inherent in the species. But whereas birds' restlessness is always aroused through hormones by an inner calendar, many butterfly species turn into migrants only in years of great over-population. As with locusts, it is a kind of psychological stress which triggers off travelling.

Insect Builders

However, it is not only in traveling that creatures can outrival man. In building, too, even recent inventions have their parallel in the animal world—including air conditioning!

The way in which some insects organize themselves in order to build their dwellings has always aroused the admiration of observers. One of the most spectacular examples is the case of a species of termite (*Macrotermes natalensis*) found on the Ivory Coast of Africa.

These termites need a tropical climate of 86° F to live in. The two million termites inhabiting a medium-sized colony do not rely on the natural tropical heat to maintain an even temperature, since this heat fluctuates a good deal. The walls of a colony are as hard as concrete and more than a foot and a half thick. In order to breathe, the termites require a huge amount of fresh air a day. This is provided by the colonies' built-in air conditioning plant consisting of air vents where the termite engineers are constantly busy narrowing or widening, closing or opening vents depending on the time of day or year, whether the temperature is too cold or too warm and whether there is too much or too little oxygen.

Monkey Power

Finally, it is appropriate to see how our "closest cousins"—the monkeys—may not only co-operate with us but take over our jobs! The pig-tailed monkeys living in Thailand and the Malayan peninsula have been skillfully trained to pick only fully ripe nuts and throw them down from the trees. Palm trees in which they find only unripe nuts are quickly abandoned and they concentrate on others where the nuts are ready to pick.

These macaque monkeys in Malaysia actually earn money for their labor—and "Atticus" in the English *Sunday Times* reports they can harvest up to eight hundred coconuts per day.

John H. Corner, Professor of Botany at Cambridge, officially employed monkeys during his period of office as director of the Municipal Gardens in Kuala Lumpur. Entered in the ledgers as "municipal employees," their job was to pick and throw down from the trees various rare botanical specimens. For each of these monkeys the owner received an annual salary of about fifty Malayan dollars, together with a free supply of rice, bananas, and hens' eggs. When the monkeys were not actively employed by the municipal administration they were hired out to managers of coconut plantations. Relations between the monkeys and their owners were very relaxed. Before starting work the monkeys were let off the lead, and after they had finished their tasks they returned of their own accord, allowing their leads to be replaced, just like well-trained dogs.

The use of chimpanzees to operate a conveyor-belt in a furniture factory in Texas was also reported by a *Sunday Times* columnist. According to the article, three chimpanzees were employed to stuff foam rubber into sofa cushions and zip them up. It was also claimed that each of the chimpanzees did the work of three women. Objections to their being employed were raised not by the animal protection society but by the upholsterers' trade union, which sent a delegation to the workshop to clarify the legal position regarding the use of animal labor! However, zoologists normally consider chimpanzees incapable of work conforming to set standards, however adept they may be at short-term imitation, mainly because of their sudden changes of mood.

ANIMALS AS SYMBOLS

A single chapter, of course, can only touch the fringes of this fascinating subject of animals as symbols. To do it justice would demand not just one book but a whole library of them. Think of all the volumes that have been written about the animal cults and the animal (or half animal) gods and goddesses of Ancient Egypt, or about all those myths of the Ancient Greeks and Romans in which gods assume animal shapes, or about the organization of many of the American Indian tribes round their various animal totems, to name just a few facets of the subject. To explore all this would mean entering the overlapping fields of anthropology, social and cultural history, and psychology.

It would be a reasonably fair simplification, though, to say that from the beginning of time man has, in one form or another, incorporated the animals with whom he shares the earth into his own religious fears, strivings and aspirations. A surprising variety of animals figure in this way. Here, in alphabetical order, are just a few examples—and you will find others in the rest of the chapter:

BEAR: The bear is revered among all primitive peoples who come into contact with it, and there are numerous festivals in its honor. The cult of the bear has been particularly strong in east Asia among the Siberian tribes. Many specialists now believe that the Ancient Greek goddess Artemis was connected with the cult of the bear: girls danced in the guise of bears in her honor, and could not marry before undergoing this ceremony.

BUFFALO: The Todas of southern India never eat the flesh of their domestic animal, the buffalo—except once a year when they sacrifice a bull calf, which is ceremonially eaten in the forest by the adult males.

CATTLE: Cattle are respected and often afforded semi-divine status, among many pastoral peoples. In Persian mythology, the worship of Mithras included the sacred bull which had to be slain to give life to the earth. The bull's soul, Sylvanus, became the guardian of all herds and flocks. Mithraic cults extended throughout the Mediterranean, and the Spanish bullfight may have connections with it. Conspicuous among the animal cults of Ancient Egypt was that of the bull, Apis. Its birthday was celebrated, oracles were obtained from it, and after death it was mummified, and buried in a rock tomb. In India respect for the cow is still wide-spread, and in many parts of the country it is forbidden to kill and eat it—an injunction strictly observed even in times of famine. The products of the cow are also important in Indian magic.

CROW: The crow is the main deity of the Thlinkit Indians of northwest America and, together with the eagle hawk, plays an important part in the mythology of southeast Australia.

ELEPHANT: The elephant is frequently regarded as a sacred animal. In Thailand the rare white elephant is believed to contain the soul of a dead person, perhaps a Buddha, and it is baptized, fêted, and at its death mourned for like a human being. In some parts of Indo-China it is believed that the soul of the elephant may injure people after death. In Cambodia the elephant is considered to bring luck, and in Sumatra it is regarded as a tutelary or guardian spirit.

FISH: Dagon, worshipped by the Ancient Babylonians and other Semitic peoples, was a fish-god, and his devotees wore fish skins. There have been other instances of fish deities (and the legend of the mermaid may derive from them).

GOAT: Dionysius, the Ancient Greek god of emotional release and of wine or spirits, was believed to take the form of a goat. Other Greek rural deities, among them Pan, Silenus, the Satyrs and the Fauns were usually part-human and part goat (though some of them had horse-like features).

HARE: In North America the tribes of Algonquin Indians had as their chief deity "a mighty hare" to whom they went after death.

HORSE: The horse figures a great deal in myth and legend. It was the winged horse, Pegasus, of course, which carried the thunderbolt of Zeus, the chief of the gods of Ancient Greece. The Centaurs were legendary as creatures having the upper parts of a human and the lower parts of a horse. Poseidon, the Greek god of earthquakes and water—and especially the sea—was usually depicted as part-horse, as were many lesser water gods. In Ancient Gaul there was a horse-goddess named Epona, and the Gonds of India worship a horse-god, Koda Pen.

LEOPARD: The cult of the leopard used to be wide-spread in West Africa, and still is in the more remote regions. Among the Ewe a man who kills a leopard is considered to have committed a great sacrilege and is in danger of being put to death. In Loango a prince's cap is put upon the head of a dead leopard and dances are held in its honor.

LION: The lion is associated with the Ancient Egyptian gods Ra and Horus. The Arabs used to have a lion-god, Yaghuth, and among the Balonda of Africa a lion-idol is sometimes still worshipped.

LIZARD: The cult of the lizard is prominent in various parts of the Pacific. It is also a guardian deity in Madagascar.

MANTIS: A praying mantis seems an unlikely candidate in this context, but among the Bushmen of Africa it is regarded as an incarnation of one of their gods (the caterpillar is another).

MONKEY: In India the monkey-god, Hanuman, is one of the most prominent figures in Hindu mythology.

SERPENTS: The cult of the serpent has been common in many parts of the world. Asclepius (or Aesculapius), the Greek hero traditionally associated with the art of healing, was often afforded divine or semi-divine status in many parts of Ancient Greece and Rome, and the serpent was regarded as his reincarnation. Twined snakes round a staff is still a symbol of the medical profession. In Africa the chief focus of snake worship used to be Dahomey, and in many other parts of Africa snakes are thought to be the incarnations of dead relatives. In the old Natchez culture of South America the rattlesnake was worshipped, and the Aztec deity Quetzalcoatl was a serpent-god. In many parts of India there are carved representations of cobras, to which food and flowers are offered.

SHEEP: Ammon, the god of Thebes, in Ancient Egypt, was represented as having the head of a ram, and his worshippers regarded the ram as sacred.

TIGER: The tiger is associated with the Hindu gods Siva and Durga, and there is a tiger cult among some of the more primitive peoples of India. In Nepal there is a tiger festival known as Bagh Jatra, and the worshippers dance disguised as tigers.

WOLF: Both Zeus and Apollo were associated with the wolf in Ancient Greek mythology. The Thlinkit Indians of northwest America had a god whose name means "wolf," and they worshipped a wolf-headed image.

Even that short list shows how closely animals of all kinds have been involved in man's attempts to make sense of the world in which he lives, and to grapple with the great mysteries of life and death. There are many possible explanations for it. It seems natural enough, for instance, that at a time when men were utterly dependent for their survival on the animals they hunted and killed, or on their domesticated animals, they should regard them with a gratitude so intense that it inevitably shaded into awe and reverence, and that they should seek to propitiate the animals and encourage them to multiply with gifts, prayers and rituals, by making drawings or images of them—that they should in effect turn them into gods or the representatives of gods. Other primitive peoples who have been vegetarian in their diets have worshipped the plants or crops upon which they depended in much the same way.

No doubt, too, primitive man realized very clearly that certain animals possessed qualities—courage, patience, cunning and so on—which he needed in his own struggle for survival, and again it seems only natural that he should express his admiration in the form of various rituals, in the hope of acquiring these qualities for himself.

In any case, at an early stage in man's development the gulf between man and the animals wasn't anywhere as wide as it has become since. For one thing, the daily struggle for existence was equally hard for both of them, and it took place in much the same habitat. Primitive man didn't (and doesn't, where he still exists) assume, as we tend to, that animals are separated from himself in any fundamental sense. They have frequently been regarded, for example, as the temporary or permanent abode of the souls of dead humans, and sometimes as the actual souls themselves. The animals concerned, therefore, would have to be treated with respect, both because the kinsmen

of the dead would want to preserve their good-will, and because they would want to spare them unnecessary suffering.

In many other cases, too, animals themselves were believed to have souls of their own, just as much as human beings. It is this belief that helps to explain the cults of dangerous animals like lions and tigers: they are in part inspired by fear that the soul of the slain animal might take vengeance on the hunter, together with a desire to placate the rest of the species. As a matter of fact, this question of the souls of animals continued to be a matter of debate well into civilized times.

As far as this chapter is concerned, the psychological factors are of particular importance. There is a sense in which we *need* the animals: they represent aspects of ourselves, realistically and satisfactorily externalized. This is one of the reasons for their perpetual fascination, and it's one of the reasons, also, why the disappearance of so many species of wild animals is dangerous to the well-being of man, quite apart from the ecological considerations. In the past, the discovery of new species has always aroused excitement among ordinary people as well as among the zoologists. Man finds it comforting to be reassured from time to time that he is not alone in his world, that there are still strange and wonderful creations in Nature—and animals help to meet this desire too. It could almost be said that if animals didn't exist, then man would have to invent them.

The Oldest Friend of Man?

Dogs have been with man for so long that often they aren't looked on as real animals but almost as honorary humans. There's that lovely old legend (or evolutionary metaphor) about the moment in prehistoric time when suddenly a chasm began to open between man on one side and the whole animal creation on the other. As the chasm grew, the dog, after a last look over his shoulder at his fellow-animals, leaped across the gap just before it became too wide, and landed at the side of man.

Then there is the heart-breaking Welsh legend of Bethgelert (which means grave of Gelert), over which countless English schoolchildren have wept. According to a tradition in the village at the foot of Mount Snowdon where Llewelyn the Great lived, Gelert was a magnificent hound given to Llewelyn by King John. The dog was greatly loved by his master, but one day it was missing when Llewelyn went out hunting. When he got back, he was horrified to find Gelert smeared with blood. Going in, he found his beloved child's bed empty and in a terrible mess, and the child nowhere to be found. Mad with grief and temporarily out of his senses, Llewelyn turned on his dog, believing him responsible, drew his sword and killed him. Wakened by the animal's dying howls, the child cried out from under a heap of coverings where it had been hiding. Snatching up the child, Llewelyn was appalled to find under the bed the body of a huge wolf which his faithful Gelert had fought and killed in protecting the child.

This legend, which speaks so movingly of dogs' devotion to man, is also to be found in many other places and it is believed that it may have come originally from India.

It was, perhaps, partly because of dogs' very familiarity and because they often shared man's roof and hearth, that though they frequently roused affection, they rarely inspired religious feelings.

There is one breed of dog, though, which has traditionally inspired a respect in man not far removed from reverence. This is the Saluki. It is a very ancient breed, and carvings and paintings have been found in Egyptian tombs dating back to 6,000 B.C. which depict Salukis almost identical in appearance to those of the present day. A number of their mummified bodies have also been found in the Tomb of Kings, and this may in fact suggest that this particular breed of dog *did* have religious attributes or associations.

The Ancient Egyptians used the Saluki for hunting wild gazelles, and the Arabs still use the dog for that purpose. This is because of its phenomenal eyesight even more than its speed. At one time it was used in conjunction with a falcon. The bird would rise to a height that made it no more than a speck to human eyes. Then, as the speck began to hover, the Saluki would be released and immediately ran to the quarry the bird had spotted. Even today's Salukis, generations removed from the old desert stock, can spy a bird or a plane in the sky before it is visible to the human eye.

The Arabs of the desert hold the Saluki in the same high regard that the Ancient Egyptians did. To the Arab, dogs in general are unclean animals, but the Saluki is an exception. It is, together with his horse, an Arab's most valued possession, sharing his tent and his food and his whole way of life. Neither dogs nor bitches are allowed to mate with any other breed, and it is often claimed that a Saluki bitch will not allow any other breed of dog to approach her. As with the best Arab horses, a Saluki's ancestry can therefore sometimes be traced back over thousands of years. It's little wonder that such animal aristocrats are treated with special respect. Even today it is said that an Arab sheikh, visiting a western country and meeting a Saluki, will often bow to the hound before shaking hands with its owner.

In the past an Arab would never sell a Saluki, but only give it as a present to a specially honored friend. It is this practice which probably explains why the breed took so long to reach the West.

Another canine aristocrat is the Afghan hound. It bears a strong resemblance to the Saluki, and it may be that at one time Salukis were taken from Egypt to Afghanistan. Or perhaps two breeds, produced for similar purposes, developed similar characteristics. Images of dogs very like the Afghan hounds of today are to be found on rock carvings in the caverns of Balkh in northeast Afghanistan, dating back to 2000 B.C. According to folklore this is the breed of dog which Noah took with him in the Ark.

There seems to be no European equivalent to the Saluki and the Afghan hound as far as prestige is concerned. However, in Europe, too, hunting dogs were often highly prized, and frequently figure in historical tradition and legend. An outstanding example is the Scottish deerhound which, like the Saluki and the Afghan hound, hunts primarily by sight rather than scent. No one seems to know how this breed of deerhound first reached Scotland or where it came from. Some authorities think it was introduced from Normandy at the time of the Conquest, but others believe it may have existed in Scotland before that. A story told by Raphael Holinshed, the sixteenth-century historian, in his *History and Description of Scotland* (itself probably an adaption from another author), seems to bear this out. It tells how a group of Pict nobles were out hunting one day with the King of the Scots, and found to their chagrin that the King's deerhounds were much better than their own. The King generously presented some of his hounds to the nobles. The best of them, though, he kept for himself, and that was the dog the Picts really coveted. So they stole the hound, and the result was a bloody battle in which, Holinshed tells, "sixty Scott gentlemen" and over a hundred Picts were killed.

There's another well-known Scottish legend of the fourteenth century which also illustrates the esteem in which the Scottish deerhound is held. Sir William St. Clare owned two fine specimens named Help and Hold. He was foolish enough to brag about them to King Robert the Bruce, who knew only too well that none of his own hounds could compare with them. Eventually in exasperation Robert issued a challenge to Sir William that Help and Hold should pull down a "white faunch deer" on Pentland Moor, not far from Edinburgh. "Pentland Moor is yours if your hounds hold the deer" he told Sir William, adding "Your head comes off if they lose her!"

Help and Hold found a white faunch deer all right, but Sir William's heart must have been in his mouth when the deer jumped into the stream which formed the boundary of the Moor. Fortunately for him, Help and Hold plunged in after her, pulled the deer down, and held it, before it reached the other side. And so Sir William kept his head and the whole of Pentland Moor as well.

A Catalogue of Feline Facts, Traditions and Legends

No one who owns—or rather, is owned by—a cat should be surprised to find out that this has always been regarded as a very superior animal indeed, even though it isn't as ancient a companion of man as the dog. It is not easy to say exactly when the cat was first domesticated. Cat bones have been found in prehistoric cave dwellings, but they were almost certainly those of wild ones, possibly killed for their meat. It is generally agreed that cats first became part of the household in Ancient Egypt, about 3000 B.C. (when agriculture was already fully established) when wild cats were first tamed and used to protect the granaries from mice and rats.

Cats soon became very highly thought of, not only for this practical part they played, but also for their own sake. It may have been in order to give *them* protection that they were made sacred animals, associated with the goddess Bastet or Pasht (Pasht, incidentally, is sometimes thought to be the origin of the pet name Puss), and therefore part of the whole divine order of Ancient Egypt. In consequence, cats were worshipped in their particular goddess's temple (as other sacred animals like baboons and crocodiles were in theirs), and on her feast days they were garlanded,

led in procession and generally treated with a reverence which—knowing cats—they probably took as their proper due! When they died, cats were buried with great pomp in their goddess's holy ground. In many cases their corpses were embalmed and mummified; in the mid-nineteenth century no less than 300,000 mummified cats were found on the site of a temple originally built in honor of Pasht.

The export of cats was strictly forbidden by the Ancient Egyptian authorities. A lucrative contraband trade in them grew up, however, mostly conducted by Phoenician and Greek traders. Although the government appointed special agents to try and break the smuggling rings, the prized monopoly in domesticated cats was lost, and gradually the animals reached many other countries. By 1000 B.C. they were well known in China, and from there spread to Japan and probably India—though some evidence suggests that the domestication of cats there had begun as early as it had in Egypt. In Japan cats were regarded as so precious that they were always kept on leads, until 1602 A.D. when the government ordered their release, so that they could deal more effectively with the vermin that threatened the silkworm industry.

The Romans adopted the cat as a symbol of liberty, and their armies took cats with them to practically every corner of Europe. In Britain, it was probably the Phoenicians who first introduced cats when they landed in Cornwall to trade for tin, long before the Roman invasions. The Romans brought cats to Britain too, as the discovery of the remains of a cat in a Roman villa at Lullingstone, Kent, proved. There were cats early on in Scotland also, where they were adopted as symbols of courage—the origin of the county name of Caithness is said to have derived from the word cat. In Wales the animal was specially protected by legislation introduced by Hywel Da, King of South Wales, in 936 A.D., which fixed the penalty for killing a cat at the equivalent of its worth (which was quite considerable) measured out in corn.

Cats didn't always appear in such an amiable light, however. There are a number of medieval and Renaissance religious paintings, for example, including several by Tintoretto and Veronese, which use the cat as a symbol of betrayal and death in dealing with various aspects of the life of Christ. Sometimes a cat is shown near Judas at the Last Supper—as in a famous fresco by Ghirlandaio in the Museo San Marco in Florence.

This perhaps was prophetic of the horrific treatment meted out to cats and humans alike during the witch-hunts of the sixteenth and seventeenth centuries. The practice of witchcraft, still widespread in many parts of Europe—as well as in the colonies of North America—was evidence of the obstinate survival of ancient pagan fertility rituals. As in the religions of Ancient Egypt various animals figured prominently in these cults—and one of these animals was the cat. In

addition there had been a revival in the Rhineland of Germany of a cult associated with the old Norse goddess Freya, whose chariot was traditionally drawn by two black cats, and in whose rituals cats played a major part. The Christian Church believed that the time had come to put an end to these pagan practices, and it set about doing so with a ferocity which to us seems as opposed to the true spirit of Christianity as it possibly could be. The cat had now come to be regarded as one of the embodiments of the Devil. Merely to own one was to risk arrest for witchcraft. Thousands of lonely old women, whose only companions were their cats, were hunted down, tortured and burned along with their pets. Stray cats and wild cats were hunted down with equal cruelty. On saints days, such as the Festival of St. John in France, sack-loads of cats were thrown onto the bonfires.

It was a long time before the fury began to die down, and the cat was once more restored to its earlier prestigious position. The cat populations in both Europe and North America had in fact been so reduced that for a time cats were hard and expensive to come by. It is said that the first cat to be imported into Paraguay in 1750 cost its weight in solid gold! Even today it is difficult to be sure whether the fear that some people have of cats is the result of a genuine physical allergy or psychological hang-up—or is instead the residue of our ancestors' primitive and irrational terrors.

For most of us, of course, our cats are loved and honored friends of the family, showing the sort of controlled arrogance which suggests they have even more ancient race-memories of the elevated place in society which cats once held—and feel they should still hold today!

Fabulous Beasts

Keith Poole

One of the specialized fields in which man's fascination with animals, both real and imaginary, has expressed itself is that of heraldry. Everyone who has wandered round the streets of European cities must have seen and wondered at those carvings and paintings of animals, some of them very odd looking, above the gateways and doors and roofs of ancient buildings. Included among them are many coats-of-arms, and especially, of course in the United Kingdom, those of the British Royal Family.

The correct name for these heraldic animals is "Beasts," and Keith Poole, who is an expert on the subject, has written here about some of them:

The Royal Beasts

As far back as the 12th century the word "Beasts" was used to define any living creature, and so included birds, fish, reptiles, and mythical monsters. Dukes, earls and barons all had their personal beasts of stone or wood, which they set up in their castles, manor houses, gardens, on the gable ends of buildings, roofs, newel posts of staircases, and chimney-pieces. They decorated their banners, standards, flags and shields; they wore them as crests on their helmets in tournaments. They were carved on their seals and fashioned in precious metals to give as presents to distinguished guests.

The commonest choice was the lion, and often a pair of stone beasts, over six feet high, flanked the castle entrance. It was not until a century later that heraldic beasts were used, known as "supporters," because they supported the nobleman's shield of arms at each side when it was engraved in wax or metal.

The heyday of the "Beast," however, was in the 15th and 16th centuries. One of the best known beasts is the bear holding a ragged staff, the badge of Warwick "the Kingmaker," used on his banners, standards, and the livery of his retainers. It would be impossible to list the many Royal Beasts chosen by successive sovereigns. They included greyhounds, lions, dragons, harts, antelopes, panthers, bulls, swans, a dun cow, boars, falcons, and eagles, to name only a very few. Henry VIII used no less than eighteen beasts to top the tent poles of his pavilion when in France. Queen Elizabeth I ordered ten wooden beasts to guard Rochester Bridge; a lion, unicorn, buck, greyhound, bull, boar, dragon, leopard, talbot (a hunting dog), and panther; each one holding a shield bearing the Royal Arms. She also had six "Beasts Royal" adorning the fountains and cisterns in Windsor Castle, and ten more on the stone fireplace in the Castle Library.

St. George's Chapel, Windsor, is full of beasts, inside and out. On the buttresses a series of Royal Beasts was set up in Tudor times denoting Henry VIII's descent from Edward III through various lines. They became so dilapidated and weather-worn that Sir

Christopher Wren, considering them dangerous, had them removed some 250 years ago. In 1925 they were replaced by seventy-five new ones.

Henry VIII outshone every monarch before or since in his passion for more and more beasts at his palace at Hampton Court. He filled it, both inside and out, with beasts of all sizes and varieties. The lion, dragon and greyhound predominated; but there were harts, yales, bulls, antelopes, and many others set up in the Great Hall, on fountain cisterns, forming avenues in the gardens, on the roofs, parapets, gable ends and bridges. The path to "The Mount" was lined with them: lions, greyhounds, dragons, harts, bulls, leopards and antelopes, all holding weather vanes. There were scores of them scattered throughout the spacious gardens which, having been planned during Anne Boleyn's time, everywhere displayed her ten leopards with crowns about their necks. After her execution the Royal accounts include a provision for the erection of four dragons, six lions, five greyhounds, five harts and four unicorns, all to be carved in freestone, the unicorns being for Anne's successor Jane Seymour. This extravagance was offset by the strange economy of cutting off the heads and tails of Anne's leopards and turning them into Jane Seymour's panthers.

Most of this remarkable collection of beasts has long ago vanished, only ten now remaining on the bridge spanning the moat, and these are modern, being erected in 1950. There were originally twelve, two lions, two dragons, two greyhounds, two unicorns, two panthers, a bull and a yale, but one unicorn and one greyhound have been dropped.

The Queen's Beasts

One of the most attractive of many decorations set up in London for the coronation of Queen Elizabeth II on June 2, 1953, was the set of ten plaster beasts outside the great window of the annex leading to Westminster Abbey. Thousands of visitors gazed at them with puzzlement, wonder and delight, and very few indeed knew that these huge six-feet-high figures, each weighing half a ton and holding a shield bearing the Royal Arms, were the Queen's Beasts. The ten were carefully chosen from the Royal Bestiary of some thirty or more belonging to the Queen through dynastic inheritance.

They had been specially designed from Henry VIII's earlier beasts at Hampton Court Palace, known as the King's Beasts, some 400 years earlier. When William

III became king he ordered all the beasts to be thrown into the moat and the moat filled in. There they lay until 1909 when the moat was cleared and various parts of the beasts were discovered, to serve as a guide for the Coronation beasts.

Outside Westminster the ten beasts stood like sentinels guarding the path the Queen would take, the place of honour rightly given to the Lion of England, royally crowned. Then came the greyhound, yale, dragon, horse, white lion of Mortimer, unicorn, griffin, bull and falcon, in that order. They were chosen not in order of pedigree but for their artistic balance and symmetry. The beasts themselves were not coloured, only the shields they supported.

After the traditional pageantry, colour, beauty and excitement of the Coronation, these splendid beasts, having perfectly fulfilled their Royal duties, were transported to the Great Hall of Hampton Court Palace, where they rest in peace. Copies of them in stone are in Kew Gardens.

The following listed order of the ten beasts is the order in which each of them entered the Royal pedigree.

1. THE LION OF ENGLAND

The lion, in earlier times called a leopard, very naturally has pride of place, being king of all the beasts. Kings and princes in Norway, Denmark, Scotland and Wales as well as England chose it for its courage, strength and ferocity. It was not, however, only a Royal Beast, for many knights had them proudly painted on their shields. Richard Lionheart was the first English king to have three gold lions on a red shield for the Royal Arms.

2. THE GRIFFIN OF EDWARD III

The griffin was a purely mythical beast believed to be over 3,000 years old and sacred to the sun. It has the head and wings of an eagle, and the body and legs of a lion. Curiously enough these were later changed, the forelegs and claws becoming those of an eagle. Its ears were pointed and, if a female, its wings were pressed back towards each other. A male griffin has no wings but spikes and tufts stick out from various parts of its body. It stood for vigilance and guardianship over a fabulous Eastern treasure, renowned for its alertness and extraordinarily keen sight. It had the speed of an eagle, the strength and courage of a lion and could never be taken alive. It could fly off with an armed knight on horseback. Only a holy man or a saint could cure a griffin's wounds, exacting for payment one of its claws, which were rare and precious. Some of these may still be seen and bought in Eastern markets, though they are actually the horns of bulls, antelopes or even the tusks of a rhinoceros. Their eggs were also highly prized and both claws and eggs were mounted in ancient drinking vessels as a supposed antidote against poison.

3. THE FALCON OF THE PLANTAGENETS

Since falconry was one of the greatest of sports in medieval times the bird became a favourite beast. Edward III greatly loved falconry, and so did his son John of Gaunt. After the Wars of the Roses Henry VII married Edward IV's daughter, Elizabeth of York and adopted all the beasts she had herself inherited, including the falcon. For his wedding his horse-trapping (covering) was richly embroidered with falcons. When Edward IV had become king he had ordered his son to take as his badge a white falcon perched on a golden fetterlock (a kind of padlock) slightly opened to show he had forced the lock and gained the Yorkist throne.

4. THE BLACK BULL OF CLARENCE, and
5. THE WHITE LION OF MORTIMER

Both these beasts were inherited by the Queen through Edward IV, the black bull descending to him through the Duke of Clarence, whose beast it was. It had golden horns and hooves and a blue tongue. Edward IV's other beasts were a falcon, a black dragon and a white wolf.

The White Lion of Mortimer, so called because it belonged to the Earldom of Mortimer, differs entirely from the English lion because of its blue tongue and claws, its silver colour, and because it is uncrowned.

6. THE YALE OF BEAUFORT

A most curious mythical beast, it was white with golden hooves, boar-like tusks, sickle-shaped horns, and its body was covered with golden spots like coins. It was said to be able to ward off any enemy attack by swivelling its horns in all directions. It is always shown with one horn pointed forwards, the other backwards. It is the most distinctive beast of the Beauforts, descending to the Queen through Lady Margaret Beaufort, mother of Henry VII. It is magnificently displayed over the main entrances of Christ's College and St John's College, Cambridge, both of which she founded.

7. THE WHITE GREYHOUND OF RICHMOND, and
8. THE RED DRAGON OF WALES

The White Greyhound, with its red tongue and its red collar with a gold ring, was another of Edward III's many beasts, and probably of all his sons. Coursing, like falconry, was a popular medieval sport. The greyhound was prized then, as it is today, for its astonishing speed and intelligence. It was a special beast of the Earldom of Richmond (Yorkshire), hence its title.

The Red Dragon, claimed by Wales, Wessex and Somerset, was the most fearsome of all the beasts, though no one had ever seen one. It had a red body with a yellow underbelly, a barbed tail, white tusks, blue tongue and claws, and breathed fire. It is not surprising, therefore, that it was supposed to be the Devil and so was slain by Saint George. It possessed amazing strength, power, and wisdom, and its emblem was borne on banners and standards all over the world.

9. THE UNICORN OF SCOTLAND

Surely the most endearing of all the beasts. It came to England from Scotland when the Tudor line ended with the death of Elizabeth I and the Stuart line began. This unique mythical beast has the head, mane and body of a horse, a goat's beard, an antelope's legs, and a long, straight, spiral horn. It is white with golden horn, hooves, three-pointed tail, collar and chain. It was immensely strong, fierce, courageous and huge; so huge and so awkward with its horn that Noah refused to let it travel in his ark but had it towed behind. The unicorn would use his horn to stir up pools of stagnant impure water so that the other animals could drink the water it had purified. Its powdered hair was said to be an antidote for poison. When James VI of Scotland became James I of England he brought the Scottish Unicorn as one of the supporters of the Royal Arms. To make sure his Unicorn should never be replaced he chained and collared it to the shield bearing the Royal Arms, where it has remained ever since.

10. THE WHITE HORSE OF HANOVER

When George I became King of England on the death of Queen Anne he put his White Horse beast in the Royal Arms. The original White Horse goes back to Saxon times, so that the White Horse of Kent is always shown rearing up, whereas the Hanoverian one is shown running. Though older than many other beasts it is the least used, and never at all as a supporter of the Royal Arms.

1. THE LION

OF

ENGLAND

2. THE GRIFFIN

OF

EDWARD III

3. THE FALCON
OF THE PLANTAGENETS

4. THE BLACK BULL
OF CLARENCE

5. THE WHITE LION
OF MORTIMER

Animals as Symbols / *199*

6. THE YALE
OF
BEAUFORT

7. THE WHITE
GREYHOUND
OF
RICHMOND

8. THE RED DRAGON
OF WALES

9. THE UNICORN
OF SCOTLAND

10. THE WHITE HORSE
OF HANOVER

ARE ANIMALS PSYCHIC?

From very ancient times strange and often decidedly spooky stories have been told, often by thoroughly reliable people, about the capacity which animals often apparently possess for communicating with each other and with humans without, as far as can be observed, the exchange of any signal. Even stranger is their ability to "know" in advance when certain events are about to take place.

All of us can probably remember occasions when our pets have startled us by seeming to understand exactly what we are thinking or feeling. It's not so difficult, as a rule, to find a natural explanation for this. Those who know animals well know how quick they are to catch a tone of voice, a change of mood, an alteration in the atmosphere and physical details of their surroundings (the bustle preparatory to going on holiday, for instance), or even a change in the bodily stance of those near them.

In cities the idea of people "talking" to animals is usually scoffed at. But observant pet owners and people who live in the country close to animals, or who work among them, take it for granted. There are hundreds of trainers, jockeys, hunters, kennel-help, shepherds, herdsmen, vets and so on who *know* that they can, in certain circumstances, speak to animals in normal human language and be understood—though, of course, it's not the words themselves which are understood, but their sounds, the actions accompanying them, the *way* they are said, and their general ambience. There are specialists who cure animals of nervous ailments by stroking and talking to them, and it has been claimed that physical illnesses have been successfully treated by the same means. Gypsies often have an uncanny knack with horses—even to the point of persuading them to leave their owners' stables without uttering a sound!—and gypsy lore

Are Animals Psychic? / 203

contains many "tips" as to how it should be done. There are many beekeepers who "talk" to their bees and claim that it makes all the difference to their behavior and the quality of the honey they produce. In the days before mechanical milking many a milkmaid was quite certain that if she talked or sang to the animal she was tending it would give a better and more copious flow of milk.

Far more mysterious are those instances when animals obviously know that someone is approaching although the humans about are not aware of it. A college student's dog always knew when his master was coming home for a holiday, often a day or two before the boy actually arrived. It's possible, of course, that he "caught" the feelings of the boy's parents and responded to the general atmosphere of anticipation— but it made no difference if the student postponed his return by a week or so, or if he arrived unexpectedly.

One man who was working in Portugal used to return home to Carcavellos, some ten miles from Lisbon, by the little local railway. He seldom caught the same train because his hours of work were irregular, but his cat was always waiting for him on the garden wall, to jump on to his shoulders as he passed under her. She never went on to the wall unless her owner was on the way; she always knew in advance when he was coming. The station was about a mile away, but perhaps she heard the train arriving— though that doesn't explain how she knew *which* train. Or perhaps she could hear the man's approaching footsteps long before they were audible to human ears.

A sixteenth-century writer noted that "dogs by their howling portend death and calamities," and early in the nineteenth century another author wrote: "a dog's distressful howling presages death especially in houses where someone is lying ill." From time immemorial this particular kind of howling has made people shudder, because they know from experience that it often *does* mean a fatality. How do the dogs know? Is it because with their incredibly sensitive noses, they can actually *smell* the physical deteriorations that precede death? It may, indeed, be that their sense of smell has already warned them that death is on the way because it has detected certain physical manifestations of the illness or disease which make it clear—to them—that it is incurable. It has often been noted, for example, that domestic animals will turn on one of their kind who is sick, although none of the humans in the vicinity are aware of it—perhaps because the animals find the sickness is obnoxious, or possibly because their instincts tell them that it may pose a threat to themselves through infection.

When the naturalist Charles Waterton died, it was reported that flights of wild birds, to whose welfare and protection in the grounds of his home he had devoted many years, accompanied the coffin across the lake and through the park to its last resting place. This does, admittedly, sound like a legend, though many local people believed in it implicitly. There are several authenticated instances of other wild creatures —including, in several cases, swarms of bees—attending the funerals of humans who had cared for them with great devotion over a long period of time— though, of course, there may be explanations which have nothing at all to do with the human associations.

It cannot, in fact, be too often emphasized that natural explanations may lie behind all these phenomena. As you will have learned from other sections of this book we are still only at the beginning of understanding the complex range of instinctual capabilities possessed by animals. Until quite recently even the power and range of such obvious senses as sight, smell, hearing, and touch were not properly appreciated, and such matters as migration in birds and insects, responses to atmospheric pressure, color, light, the positions of the sun and moon, and the earth's magnetic fields still contain many mysteries which are currently being investigated.

Nevertheless the *possibility* remains that some aspects of animal behavior cannot be satisfactorily explained by reference to the physical senses, and there are some scientists who believe that it is only Extra Sensory Perception—ESP for short—that can account for them. The study of ESP in animals has been labelled *anpsi*—short for *an*imal *psi* (this is a term used for psychic powers—"extrasensorimotor exchange with the environment," to use the scientific jargon). *Anpsi* began in the Soviet Union in the early 1920's, when a Russian neurophysiologist, Dr. W. Bechterev, published an account of experiments which in his view supported the theory that animals are susceptible to telepathy. What Dr. Bechterev and his assistants had done (though this is a very simplified account of experiments conducted in a proper scientific manner) was to place various objects round the laboratory and to concentrate on one of them, without looking at it, touching it or indicating it in any way. A fox terrier named Pikki, who was the subject of the experiments, brought the correct object to them in a very large (and so statistically "significant") number of cases.

Not long after, Dr. J. B. Rhine and his wife (the leading investigators in the U.S.A. of paranormal phenomena), investigated a mare, named Lady, who its owner claimed, could answer simple questions addressed to her by pointing with her hoof to lettered blocks. This sounds suspiciously like "Lord" George Sanger's circus hoax with the "learned pig" which you will have read about earlier in this book—but the Rhines are reputable scientists, and they came to the conclusion, after a series of experiments, that the only way in which Lady could have come up with the right answers so often was by responding to her mistress's thoughts—by telepathy, that`is.

The Rhines and other investigators were very well aware, however, that special difficulties lay in the way of testing animals. For one thing, their extraordinarily developed physical senses might get in the way, without the human experimenters being aware of it. For instance, the object which the animal had to select might be impregnated with smells which it particularly liked, or knew the testers liked, and it's even possible that a human being who wants something may convey his wish through some special odor, left on the object when he handled it. Or again, the animal might be aware, with its senses, of all kinds of other indications of the response that would most please its owner or any other human who was paying it attention —such as the focussing of the pupils of the eyes, the flicker of eyelids, the tilt of the head or body in the direction desired by the human experimenter or even changes in his breathing and heart-beat as the animal drew near the desired object and so on. They are just tiny, unconscious indications as far as humans are concerned, but they're highly significant to the animal.

In order to counteract all of these possibilities, a machine was devised in France a few years ago which would set up the "targets" at random, without the presence of any human experimenter, and then record the animal's performance. It is claimed that the results obtained *do* suggest that some animals possess extra sensory powers. From 1950 onwards the Rhines, at their Laboratory of Parapsychology at Duke University, continued their investigations into hitherto unexplained aspects of animal behavior which might suggest the presence of *anpsi*, and were also working on ways of improving the testing of animals. In addition, they and their staff began to collect and collate thousands of reports from people all over the world who claimed to have had experiences with animals which had startled and puzzled them. Many of these reports, of course, were unreliable, exaggerated, or susceptible to straightforward scientific explanation. But a residue remained for which there seemed to be no known explanation. These were carefully investigated, and it was decided that there were six categories of experiences with animals which cropped up most frequently and which suggested that some kind of paranormal capability was involved.

These categories included the kind of experiences we have already mentioned. There were, for example, several stories among those collected by the investigators which further illustrate the power apparently possessed by some pets of knowing when their master or mistress is approaching—by means which, at present, look as if they might be "psychic." Here are two of the stories:

Man's Best Friend

A mining apprentice near Durham owned a bull terrier bitch, which he had bought as a six-week-old pup. When the dog was just over a year old her master took to going away for the weekends, leaving her in the care of his landlady. He would leave on Saturday, and return on Sunday or Monday, though the actual times of his arrival varied considerably. One Saturday he left as usual, leaving the dog with his landlady and telling her to expect him back on Monday evening. A heavy snowfall in the area he was visiting prompted him, however, to return on Sunday instead. He was unable to let his landlady know, but when he entered the house on Sunday evening, his supper was waiting for him. His landlady wasn't in the least surprised to see him, explaining that she always knew when he was on the way because his dog would take up her position by the front door about half an hour before he arrived, no matter what time that might be.

And here's the second story, which apparently suggests that a dog can anticipate his master's return after a much longer absence:

Danny, a dog belonging to a family at St. Angelo, Texas, always waited at a certain spot in the road to welcome the youngest member of the family on his way home from school. One summer, during the holidays, the boy went to a relative's farm some 300 miles away to help with the harvesting. He was away for most of the summer and rarely wrote home, and he gave no indication as to the date or time of his return. One afternoon in August, Danny went to his usual spot on the road, and no one could budge him. A few hours later the boy came walking along the road, case in hand. He had accepted an unexpected offer of a lift home.

Here, too, are two stories about animals' apparent awareness of impending disaster:

Disaster Forecasters

A lady in New York had a bulldog named Bob. Usually when her husband went to work, Bob would accompany him. One morning, however, nothing the husband could do would coax the dog to leave the house, so he left without him. An hour later Bob's mistress, who was in the kitchen washing the dishes, heard Bob barking furiously in the room next door. When she went in to see what had happened, she found Bob confronting a large and villainous looking intruder. While she was telephoning the police the man managed to get away, with Bob close on his heels. Then, his job done, the dog settled down to a peaceful sleep.

The second of the stories is about a cat and her kittens:

Several days before the earthquake of 1925 struck central California, a woman living on a farm near Santa Barbara noticed her cat moving her kittens from under a manger in the barn, where she had previously been quite content, to a rickety-looking box nearby which was about five feet above ground level. It was in the Santa Barbara area that the earthquake, when it came, did most damage—and the barn was flooded by a mass of water. The kittens, safely above water level, were unharmed.

In this example, it's true, the explanation may be that, as recent research has suggested, animals may be physically aware of the changes in the earth's magnetic fields which probably precede an earthquake, but which human beings do not sense.

Other stories collected by Dr. Rhine's team related to cases of apparent telepathic communication between humans and animals—including instances in which the animal appeared to "read" the intention of its owner to give it some unaccustomed treat. I myself have noticed that our cats often turn up in the kitchen when we have decided to give them some treat, but before we have even thought of opening the fridge to get it out.

Another category concerned animals finding their way home, without any apparent aid from the senses. Here is one of the most striking episodes:

Homeward Bound

A family living in Oregon went on a long trip east, taking their big collie Bobbie with them. In Indiana Bobbie somehow got lost. The family held up their trip for several days while they searched for him and advertised the loss in the newspapers. When they had no success they concluded that Bobbie must have been killed and continued their journey, staying for some time in Florida before returning to their home in Oregon. Six months later Bobbie turned up, his feet so badly cut and bruised that it was three days before he could walk again. The case was publicized and various people who had befriended Bobbie on his long trek wrote to his owners, so that it was possible to roughly map out the route he had taken. It was not the route which he had originally travelled with his owners, so that there was no question of his having found his way by landmarks he had noted or by any scent traces that had been left.

Even more puzzling are those cases where an animal, separated from a human or a mate to whom it is attached, has followed the object of its affection over hundreds of miles and into completely unfamiliar territory. Here is an astonishing instance from Dr. Rhine's dossier:

The "W" family living in northern California moved to a new home at Gage, Oklahoma, 1,500 miles away. Before leaving they gave their cat, Sugar, to some friends who already had one of his litter-mates. Two weeks later Sugar disappeared, but his former owners were not informed. Fourteen months after that, Mrs. W. was standing near an open window in the barn of her new home, when a cat suddenly jumped onto her shoulder, purring and rubbing itself against her neck. At first Mrs. W. and her husband thought it was some stray, but the cat made it quite clear that he knew *them*, and paid special attention to Mrs. W. Then they discovered a horny protuberance on the animal's lower back and realized that it really must be Sugar, who had had this same deformity when they first acquired him as a kitten. A little later the friends from California came on a visit and were astonished to see Sugar, whom they identified at once, and related how he had left their house over a year ago.

How had Sugar managed to find his way to his original owners? It may be that one day some perfectly normal explanation will be discovered for the ability possessed by Sugar and other animals to travel over hundreds of miles and many months—and in the absence of any sensory trail—in order to reach their owners. In the meantime, however, there are a number of scientists who believe that some extrasensory capability must be involved.

So animals *may* be psychic! And if you have any experiences that might bear this out, perhaps you would send them to the publishers of this book.

Horses and Telepathy

Henry Blake

Henry Blake is certain that he himself is able to communicate with some of his animals by means of telepathy. He and his wife tried out this experiment with one of their favorite horses, named Cork Beg.

We placed one bucket six yards to the left of the centre point, and the other four yards to the right of it, so that when Cork Beg came out of his stable there was no bias to go over to either bucket, and without interference he did go to them on a roughly fifty-fifty basis. Now we were ready to train him to answer my telepathic commands, and this proved comparatively easy to do, if somewhat time-consuming. Each morning when I fed him, I would fill one or the other of the buckets, then I would wait until I was absolutely sure I was in telepathic communication with him, and mentally visualise the bucket that contained the food. Having done this I would let him out. Within a few days he was going straight to the bucket I directed him to, and I persevered with this for a fortnight. Cork Beg, being a very intelligent animal, quickly learned that the bucket I was telling him to go to was the one that contained his breakfast.

Of course this experiment also involved a certain amount of training for me, since I had to train myself to use my will to focus my whole mind on the mental picture of the bucket, allowing nothing to distract me. And I also had to make quite sure that when he came out in the morning I was entirely in tune with him. But having made these preparations the experiment itself was extremely simple. For the first five mornings I directed him alternately to the left and the right. On the sixth morning, to make absolutely certain that he was not taking them turn and turn about by habit, I directed him again to the container on the left. On the seventh morning I directed him to the container on the left, and on the eighth morning I directed him to the container on the left. That is, for four mornings I directed him to the container to which he had a natural bias. The ninth morning brought the most difficult experiment of all. For four mornings running he had taken his breakfast from the container on the left, and on the ninth I wanted to change him to the container on the right. Much to my relief he went straight to it. Having come out of that successfully, he had to take it again from the right-hand container on the tenth morning, from the left on the eleventh and on the twelfth morning from the right. Each morning he went directly to the correct container.

From: *Talking With Horses*, by Henry Blake.

Do Animals See Ghosts?

Many people have had the experience of seeing their pets behave in an odd and decidedly spooky fashion, and for no apparent reason. In most cases it's a dog that's involved. The animal will suddenly get up and, fixing its eyes on what is apparently empty space, will begin to whine, growl or bark, and sometimes turn and flee, tail between its legs, with every appearance of terror. This kind of behavior might merely be the result of a remembered nightmare or even of the revival of some racial memory-trace going back to the far distant times when the animal's wild ancestors constantly faced danger and terror. There are many, however, who believe that the animal's curious behavior is due to its awareness of another dimension, which humans are unable to see or feel, perhaps because *their* senses are too weak and crude to do so.

In many cases these animal responses are associated with sites which have, sometimes for centuries, traditionally been regarded as haunted. Here are some examples.

The first of them concerns a small hotel at Portreath, Cornwall, England. It was originally built as a fisherman's cottage in the sixteenth century, and later on became a private residence for many years. It had long been regarded as haunted by the ghost of a small young man dressed in eighteenth-century style who in life was presumed to have been a smuggler. The ghost was said to appear, at irregular intervals, from the wall panelling in a first floor corridor, to walk in a furtive fashion towards the staircase—and then to vanish. The spot at which he allegedly appears is the old entrance to a tunnel leading down to the beach. The tunnel was sealed off by the hotel owners in order to prevent any unauthorized person getting into the hotel.

About twenty years ago some alterations were made to the building and a secret "cubbyhole" was discovered between the floors adjoining the central staircase. In it was found the skeleton of a man, with a few remnants of an old black cloak still clinging to it, seated at a small table. There was also an old sea chest containing a few coins of the 1700's, some scraps of cloth, and an old rusty sword. The sword and the table were presented to Exeter Museum, where they can still be seen.

This discovery didn't affect the hauntings, however. Not long after a group of people, including an experienced "ghost hunter" and accompanied by a Great Dane, kept a nighttime vigil on the landing. At 1:30 in the morning the watchers felt a sudden drop in the temperature—and the big dog, who was lying beside them, tensed, his coat bristling, snarling and growling at the wall panelling that covered the entrance to the sealed tunnel. Then he got to his feet and with hackles raised and teeth bared, backed against the stairway, turning his head slowly as if he could see some presence passing close to him. He stood growling savagely for a few seconds and then, just as suddenly, lay back down again.

"The Royal Oak," a very old inn in the village of Langstone in Hampshire, England, is also haunted at intervals—though these are sometimes of considerable duration—by a woman in white. Phantom footsteps are also heard, and sometimes a scraping sound as if someone upstairs were pushing chairs around. The landlady's dog was particularly affected, and refused to be left alone in the bar.

A private house in Hampshire, England, also has a ghost. It is that of a coachman who, some 200 years before, murdered his rival in love, the butler of the household. One of the most recent investigators found that his Alsatian dog refused to enter the room where the murder is said to have taken place. On one occasion, when he tried to force the dog to enter, it cowered back, whining, and bit his master in protest, although normally it is the most docile of pets.

Cats, too, it seems, sometimes have the same psychic sensitivity. There's an old rectory in a village a few miles north of Blandford Forum, Dorset, England, whose garden was traditionally haunted by an old gardener and a lady dressed in a purple dress of the Tudor period, and sometimes by a pair of slippers moving up and down the stone steps. These manifestations had stopped for many years, until in March 1971 the vicar's wife found that there was a certain spot in the garden which neither her dog nor her cat would approach. If she tried to carry them to it, they would struggle desperately to get away. Visitors also complained that if they walked over this piece of ground they experienced a sensation of extreme cold.

But here's an example which shows that these baffling reactions aren't confined to dogs and cats. Inverawe House, in a remote part of Argyllshire, Scotland, and home of one of the Campbell families for centuries, has one of the most well-attested ghosts in the British Isles. It is of a young girl, known locally as "Green Jean" because of the green dress she is always wearing. As a rule she is only seen inside the house— but on one occasion one of the family's shepherds was driving a flock of sheep along the Inverawe road. Suddenly the sheep stopped and divided in order to let "something" pass through their ranks—and a moment later the shepherd himself saw the apparition of "Green Jean."

But can animals themselves be ghosts? They certainly figure quite often in folklore as Keith Poole recounts in two spine-tingling examples.

The Spectral Ape

For five centuries Athelhampton Hall in Dorset, England has been haunted by no less than six ghosts, the only animal one being an ape. It is called Martyn's Ape after the first owner of the house whose motto was: "He who looks at Martyn's Ape, Martyn's Ape shall look at him." This ape particularly haunts a small room leading down from the Long Gallery which has a secret staircase to the Great Chamber below. In that room one of the Martyn daughters is supposed to have committed suicide after being jilted by her lover. For a long time she had had an ape for a pet who never left her side, night or day. In her sorrow and desperation she had not noticed the ape had followed her into the room which she locked and bolted before taking her life. The ape, no doubt frantic with hunger, thirst and fear at being trapped with his lifeless mistress, itself died beside her, both skeletons being later found. The ape has ever since haunted the room, the saddest of all the ghosts in the Hall.

The Ghost Bears

Perhaps the rarest animal ghosts to be seen in England are bears. Since bear-baiting was once one of England's most popular cruel pastimes this is surprising. One of the more terrifying stories of a bear ghost took place in the Tower of London. It was January 1816 when a sentry was posted at midnight outside the Jewel House. Since there were a number of ghosts in the Tower sentry duty was never an enviable job, especially around midnight. According to a sworn statement of the incident later given by the Keeper of the Crown Jewels himself, this sentry was a particularly happy and brave man who had been singing and whistling in the guard room before his duty began.

He had only just started his patrol when, to his alarm, he saw a figure resembling a huge bear coming from under the Jewel Room door. He challenged it and as it made no answer thrust at it with his bayonet so forcefully that it stuck in the door. Then, as the bear still came lumbering towards him, he fell to the ground, senseless with fear.

Swifte, the keeper, saw the sentry the next morning, together with his fellow sentry, who testified to the former's alertness and heard his challenge before he collapsed. The following day Swifte found the sentry changed beyond all recognition. Two days later, without ever uttering a word of his ordeal, he died and was buried in the churchyard of the Tower. No explanation of the bear was given, but the guards were doubled from that day.

Strangely enough, an almost identical incident took place at Worcester during the English Civil Wars. A bear was said to haunt the precincts of the cathedral but had never been seen until it appeared one night to a sentry on guard duty. It brought such terror to him that he dropped his rifle and fled. The next day he was ordered to be court-martialled for desertion of his post in time of war.

THE EXTINCTION CRISIS

In reading this book you will have noted that the question of the survival of some of the creatures dealt with kept cropping up. It would, indeed, be impossible —and irresponsible—not to pay some attention to the question of extinction in a book like this.

There's nothing new, of course, about the dying out of certain species. It is a natural part of evolution, as the fossil record shows us. Most of us probably know, for example, that the dinosaurs (in all their varieties of forms, habits and sizes) were for more than a hundred million years the "lords of creation," probably with no other creature bigger than a cat to challenge them, during most of that immense period. Then, amazingly, they suddenly became extinct, for reasons which are still only partially understood. Man certainly had nothing to do with it, because he didn't exist. It was only with *his* advent that the possibility of any serious interference with the "natural" rate of

extinction among other species arose, and not then, of course, until he was sufficiently numerous, intelligent, and technologically advanced to become the most efficient of all the predators. It wasn't, in fact, until the seventeenth century that the kind of threat posed by man to other species really became apparent. That's when those industrial processes began which were eventually to transform man's social and economic life and, along with it, the natural environment itself.

At first this threat was greatest for those species which were in any case rare or vulnerable. In 1681, for example, the last of the Dodos died. This may have happened in any case, because the Dodo (a large flightless bird about the size of a swan) had a comparatively narrow territorial range, being confined to the large islands of Mauritius and Réunion in the Indian Ocean. But it was the expansion of the human populations of the islands, together with the intro-

The Extinction Crisis / 213

duction of dogs and pigs that sealed the creature's fate —for the Dodo laid its single large white egg on a bed of grass on the floor of the forests, where it could only too easily be disturbed. Of the 85 other species of birds which have disappeared from the face of the earth since then, naturalists have estimated that about a quarter of them might have become extinct naturally, while the rest have been the victims of human activity in some form or other. The figures for the mammals are roughly comparable, and taking birds and mammals together, since the year 1600 man has been responsible for the obliteration of about 75 per cent of the vanished species. That is, the rate of extinction has been three to four times greater than it would have been in the natural course of events.

The worst period was between about 1890 and 1910, when industrial and imperial expansion and competition among the "advanced" nations were at their height. Many governments then realized, at least to some small extent, what was happening to the wildlife around them, and introduced legislation limiting the slaughter of some endangered species and providing protection for others, and there was a considerable increase in the number of wildlife sanctuaries and preserves in various parts of the world. Between 1910 and 1960, in consequence, the rate of loss among both birds and mammals was appreciably reduced (though the commercial exploitation and decimation of the whales continued unabated).

But after 1960 a new and far more serious crisis loomed up, not so much because of man's commercial depredations (though these were serious enough), but preeminently for environmental reasons linked to the population explosion. The new threat can be seen as a two-fold one. First and most obvious, there is the inevitable shrinking of the natural habitat as more and more land has to be brought under cultivation in order to feed the extra millions of hungry mouths, especially in the developing countries, and above all perhaps in South America where the clearing of vast areas of primeval forest means the end of many wildlife species. Secondly, and even more serious, there is the ever increasing spread of new industrial technologies, and the frantic search for natural resources to sustain them. The chlorinated hydrocarbons and other residual poisons produced by these advanced industrial processes and technologies have wrought such havoc with the natural environment in many parts of the world that the plight of many wild species is now worse than it has ever been. As James Fisher, the much-decorated ornithologist and conservationist, has said: "Of all the animals man is the dirtiest. He is the most powerful pollutant, destroyer, eroder and exploiter that the biosphere has ever encountered"— in spite of the fact that man himself is, of course, a product of that biosphere, and inevitably belongs to it.

To try and meet the renewed threat to the world's wildlife, private, national and international organizations have sprung up. The most important of the international ones are the International Union for the Conservation of Nature and Natural Resources and the International Council for Bird Preservation. The first of these (IUCN for short) has several separate "commissions." One of these is the Survival Service Commission, whose chairman is the distinguished naturalist Sir Peter Scott. It issues its own "little red book"—the so-called *Red Data Book*, which periodically lists both the wild animals and plants which, in all parts of the world, are in danger of extinction. It sets out to interpret the words "in danger" as flexibly as possible, so that steps can be taken to conserve the threatened species before the point of no return has been reached. But there are some species which have already become so depleted that it is touch-and-go whether they will survive. Among these are the tiger, the cheetah, the polar bear, the European bison, the swamp deer, the golden lion marmoset, the Javan rhinoceros, the fin whale, the whooping crane, the Spanish imperial eagle, the trumpeter swan, the green turtle, and several others (including some of the fishes), amounting to about seventeen in all. Some idea of the seriousness of the situation as far as this "special danger list" is concerned can be gathered from the case of the Javan rhinoceros: only one species of rhinoceros, in fact, has a population over 200—but there are probably fewer than 50 Javan rhinoceros still remaining, though it is still hoped that the situation will reverse.

In 1961, Sir Peter Scott, with the help of H.R.H. the Duke of Edinburgh, also launched the World Wildlife Fund as a kind of financial arm to the other international organizations, and in order to encourage and help the purchase of large tracts of wilderness so that they could be turned into nature reserves and national parks. It has been increasingly recognized that this is the best procedure because the life of any one wild species is indissolubly bound up with that of all the others around it, as well as with that of the plants, the soil and so on, and the preservation of a whole habitat intact ensures the continuation of the natural order.

These efforts have met with a considerable degree of success, and in many cases the rate of extinction has been held back: but it is a razor-edge kind of success, with many species just saved from total extinction, but with others becoming scarcer yearly so that the rate might suddenly begin to climb again. There is little sign of any really appreciable reduction in the amounts of poisonous substances of one kind or another which man is discharging into seas, lakes, rivers, the soil and the atmosphere upon which not only animals and plants are dependent but mankind as well.

It cannot be emphasized too often, in fact, that conservationists don't only want to save endangered species because they like them or because they are scientifically interesting (though both those motives come into it, of course) but also because we are all together on "spaceship Earth." As Sir Peter Scott has said: "What we are talking about is, in the long run, nothing less than the survival of our species. Our survival is bound up with the survival of the other animals and of the plants which, with us, make up the spaceship's crew."

The Great Flamingo Rescue

Katharine Drake

Fortunately human beings are also capable of understanding and foresight and the conservationists have their triumphs. Here to fittingly conclude our book is one such triumph, not with a well-known, "cuddly" animal but with, of all creatures, flamingoes.

In northern Tanzania, some 100 miles south of the Equator as it crosses East Africa, is an inaccessible lake called Natron. It is a strange soda lake, surrounded by steaming alkaline mud flats and quicksands; and for as long as anybody remembers, it has been the nesting ground of that strutting, crane-necked, stilt-legged, rose-plumaged bird called the lesser flamingo.

Several years ago an unprecedented drought dried up Lake Natron and killed half a million of the birds, perhaps a tenth of the world total, by cutting off their food supply. The following year floods buried the lake's mud flats under 5 feet of water, making nesting impossible. Driven by a compulsion to breed, the huge birds rose in a flaming cloud and flapped off in quest of other comparable nesting grounds. Weeks later, near exhaustion after their search, they finally dropped down—35 miles north of Natron, in Kenya—on an even stranger lake called Magadi. There disaster awaited them.

Lake Magadi has neither inlet nor outlet. It lies in a sprawl of volcanic rubble. Its sub-surface, nourished by piping-hot sulphur springs, gurgles and churns like a witch's cauldron, while above lies a motionless soda crust produced by evaporation—one vast, blinding deck two miles wide, 18 miles long, ranging in thickness from a mere film to 12 feet. So foul-smelling and noxious is Magadi that its wildlife consists mainly of vultures and rodents. Though only 510 miles south of the city of Nairobi, the area is uninhabited save for a chemical works, dredging mineral salts for export.

No one in Kenya was more delighted at first at the arrival of the lesser flamingoes—some 2 million of them—than a young British naturalist and wildlife photographer, Alan Root, and his wife, Joan. For little is known of the lesser flamingoes' way of life. Almost all that was known about their breeding requirements was that the birds need privacy—the grimmer the heat and the wasteland, the better they like it; the alkalinity of the hot soda slush from which they build their nests must be neither too strong nor too weak. Surrounding mud flats are a necessity for fledglings to walk on; so are wide stretches of open water, for the blue-green algae that floats on most of the soda lakes is the principal food of the flamingoes.

Here at Magadi was a spectacle never before witnessed by human beings: a colony actually in the process of breeding. More than a square mile of Magadi's shimmering crust boiled with their motion. Wherever the Roots turned their cameras, amorous birds swooped, soared, glided and flapped, belting out their noisy litanies of love, which are part honk, part squawk, part siren.

The nests alone were a revelation: acre upon acre of glittering, turret-shaped dwellings built up of the hot alkaline slush and baked dry in the sun, each about 18 inches high and from 12 to 24 inches across. They were jammed together, with as many as six and seven per square yard in the main concentrations. Later, when the chicks began to arrive—silvery bundles of fluff, equipped like small toys with pink beaks, goggle eyes and absurd squeaks—they stayed only on these turret tops a day or two, then with a hop and a skip they scuttled away in search of a pool.

It was near sunset on September 16, 1962, when Joan Root spotted the first hint of oncoming tragedy. Head down in a shallow lagoon lay one of the chicks, dead, legs askew. Spread-eagled beside it lay another chick, barely alive, its still unformed beak ineffectually pecking at legs that seemed encrusted in plaster of Paris. Picking it up, Joan saw to her horror that the fledgling was sightless: soda splashes had dried over its eyes like a mask. A quick look round revealed more victims, all of them youngsters. Their frail, matchstick legs, wading through Magadi's supersaturated green slime, had picked up increasing deposits which the sun baked into crippling casts.

Before dusk the Roots, desperately trying to render first aid, rounded up 100 such casualties. These victims represented only the earliest hatches. Soon baby flamingoes would burst from their shells by the hundreds of thousands. Was there no way to stave off catastrophe?

It was well after midnight when John Williams, the ornithologist of Nairobi's Coryndon Museum, was awakened by two young people carrying in their arms some small, soda-frosted baby flamingoes. There was a conference in Williams' study; Tony Irwin, head of the East African Wildlife Society and editor of *Africana* (a periodical crusading for game conservation), joined in. The night hours produced just one lead: the shackles, they discovered, were soluble in fresh water; they could be washed off. The Magadi Soda Company, reached by radio-telephone, promptly offered to make available its freshwater tanks for a rescue attempt. It lacked a pipeline, however, to channel the water from the works to the nesting areas. By breakfast, the planners had drawn up a list of equipment needed for an enormous chick-washing operation. The minimum cost would be at least £1,500.

Irwin grabbed the list and made for Nairobi's radio station. The story was broadcast out over jungle, mountain and bush. Within hours, collection boxes appeared in hotels, shops, cinemas, bars, even at street corners. Signs urged: "Make a Splash—Help to Wash a Flamingo!" By the end of the first day the fund totalled £100.

Few Kenyans reckoned on help from the outside. But on Tuesday the heartrending scenes at Magadi were shown on television screens in Britain, and a day later newspapers and radios gave the news to Western Europe, Australia and Asia. By midday on Thursday contributions and communications were pouring in from all over the world. With supplies loaded on a lorry and two Land-Rovers, Alan and Joan Root headed back for Magadi. For their preliminary task force, from the many who volunteered, they chose only two: Douglas Wise and Christopher Callow, British students on holiday. A more massive invasion, they felt, might scare away the parent birds and make them desert their chicks.

At Magadi that evening the four were greeted by a nightmare. Thousands more chicks lurched around on cruelly shackled legs. Hyenas, jackals, vultures and marabou storks were all over the place, gorging on the dead and the dying. Everywhere sounded cries of distress. Parent birds frantically flapped wings and brandished bills, striving to create a diversion. Perhaps half the stupefied adult birds still clung to their turret tops, prompted by instinct to finish the hatching. There was only one cheering note: a dozen African youngsters, children of Magadi Soda Company employees, were standing by to help round up the crippled fledglings for bathing.

On Friday, Operation Flamingo's first day, mishap piled upon mishap. The temperature rose above 130° F. The 500-yard hosepipe the Roots had brought turned out to be 200 yards short of the hardest-hit areas, meaning that chicks had to be carried to the watering terminal in canvas hampers, 40 at a time. The lake crust, deceptively sturdy to the eye, masked huge sub-surface holes gouged out by dredgers; by noon each team member had fallen through at least once, landing chin-deep in blistering-hot caustic. The terrified chicks squirmed, kicked and trampled one another in the hampers; at least half of each batch was dead on arrival. The work of dissolving casts and cleansing the soda-caked eyes and abrasions consumed more time than expected—five minutes per chick. Once free, some of the chicks scampered thankfully off; others simply collapsed and perished. By the end of the day only 200 chicks had been successfully liberated, and their rescuers were raw from soda burns.

The day had produced one ray of encouragement, however: discovery that the supersaturated, cast-forming seepage occurred only in the lake's north-eastern area, the densest area of nests. If other chicks could be kept from straying into the death-dealing lagoons, perhaps the number of prospective victims in need of wash-and-rescue might be reduced from hundreds of thousands to as few as 35,000.

That night Douglas Wise, a veterinary student, hit on a new approach to Operation Flamingo. If, in a hospital, plaster casts can be cracked off a child's leg without damage, why couldn't the soda casts be cracked off the legs of the baby flamingoes? The next morning, while Joan held a chick, Douglas gripped the cast and dealt it a quick hammer blow. The soda split in two. Another tap, and away sailed the other cast. With an incredulous headshake, the patient scampered away. The four rescuers were soon liberating a bird every eight seconds, while rubber-booted children nimbly shooed others away from the poison pools.

From then on, the ratio of chicks freed to new casualties improved steadily. Since results now warranted it, the British army sent in six 12-man teams made up of African and European volunteers. In the fierce heat and stench the soldiers tackled the bizarre operation with gusto. At the end of two weeks the score read: unshackled, 20,000 chicks; shooed away from danger, 70,000; victims still remaining, 15,000.

Meanwhile, with insufficient algae in Lake Magadi, the adult birds were faced with the arduous necessity of commuting twice daily to Lake Natron, 35 miles away, to fill their crops with food to keep their off-spring alive. The sight of the airlift never failed to revive the perspiring rescuers; determination redoubled as the vast "V" formations soared skywards, necks and legs ramrod-straight, wings indomitably beating, throats crying out a reassuring "gurook-gurook" to inform left-behind youngsters that food was on its way.

A week after hatching, the chicks had lost their innocent, Easter-card look: matchstick legs thickened, beaks began turning downward, silver fluff gave way to dark, bristly hair. Mob-minded now, the youngsters began gathering in clusters, then in crowds. In a week or so whole armies, 40,000–50,000 strong were pressed body to body. Soon, supervised by a handful of adults, the milling squawking hordes began to move like a cloud shadow towards open water, 5 miles to the south. There the youngsters would soon learn how to fend for themselves.

It was this lemming-like action that called Operation Flamingo to a standstill at the end of its 21st day—a heart-breaking decision, since some 5,000 chicks still remained shackled. But among these clamorous mobs, each new retrieve involved stampeding and endangering thousands of able youngsters. However, the ambitious mass rescue had succeeded beyond expectation. No fewer than 400,000 fledglings had survived the Magadi ordeal, 130,000 due entirely to human effort (30,000 unfettered, 100,000 saved from contamination).

The work had been punishing beyond belief. That last evening, as the Roots lay on their camp-beds, they were scarcely able to see through sun-swollen eyelids. Each was a stone lighter; they were incessantly sick; their skin was a lobster-red patchwork of blisters and sores, scars from which remain to this day. Yet exhaustion was not what was warding off sleep; neither could banish the thought of those 5,000 doomed fledglings. They could still hear piteous cries of distress rising above the clamour of parent birds, the flapping of vultures' wings, and now and again the faraway roar of a lion.

Then another note crept into the bush symphony: an almost inaudible plop, light as a whisper on the tent-top. Then slowly another plop, another and another. The Roots started up, disbelieving. They leapt outside, necks craned back, palms outstretched, staring in amazement. Rain! Marvellous, wonder-working rain! In another minute sheer cloudbursts hit the lakeside—Kenya's autumn rains, ten days ahead of schedule. Alan and Joan, soaking wet, stood spellbound.

Within 26 hours not a chick remained shackled on

the whole of Magadi. All the poison lagoons were diluted beyond further hazard.

After a rest, the Roots and the army put in two more weeks banding 8,000 fledglings, so that other secrets of the lesser flamingo's life might become known. Thanks to worldwide generosity, enough money remained to pay for the bands. Operation Flamingo itself worked out at a penny per chick.

Last year the flamingoes returned to the ancestral nesting grounds on Lake Natron, their need for human assistance over. But skylarking around East Africa today are some 400,000 exotic, rose-plumaged birds offering proof that people everywhere share a sense of compassion for Nature's creatures.

From: *The Amazing World of Nature*, by Katharine Drake.

Index

Credits

A Skunk in the Family, pp. 11-13.
Reprinted by permission from *A Skunk in the House*, by Constance Taber Colby. Copyright © 1973 by Constance Taber Colby. J.B. Lippincott Co. U.K. edition: *A Skunk in the Family*. Victor Gollancz Ltd. 1975.

A Sloth in the Family, p. 14.
Reprinted by permission from *A Sloth in the Family*, by Herman Tirler. Copyright © 1966 Harvill Press.

Bears in the Family, pp. 15-16.
Reprinted by permission from *Bears in the Family*, by Peter Krott. Copyright © 1963 by Peter Krott. Oliver and Boyd.

A Hare about the House, pp. 17-19.
Reprinted by permission from *A Hare about the House*, by Cecil S. Webb. Copyright © 1955. Hutchinson Publishing Group Ltd.

Otter Nonsense, pp. 20-22.
Reprinted by permission from *Ring of Bright Water*, by Gavin Maxwell. Copyright © 1960 by Gavin Maxwell. U.S. permission courtesy E.P. Dutton. U.K. edition: Longman Books. Penguin Books 1974.

A Gaggle of Odd Pets, pp. 23-24.
Reprinted by permission from *Starting from Scratch*, by Jeanette Travers. Copyright © 1975 Arlington Books Publishers Ltd. U.S. edition: Taplinger Publishing Co., Inc. 1976.

Mr. Worley and His Pigs, pp. 25-27.
Reprinted by permission from *All Creatures Great and Small*, by James Herriot. Copyright © 1972 by James Herriot. Michael Joseph Ltd. U.S. edition: St. Martin's Press, Inc. Bantam Books.

Unusual Patients, pp. 28-30.
Reprinted by permission from *The Animals Came in One by One*, by Buster Lloyd-Jones. Copyright © 1966 Secker and Warburg Ltd. Fontana 1968.

Winged Pets: Owned by an Eagle, pp. 31-32.
Reprinted by permission from *Owned by an Eagle*, by Gerald Summers. Copyright © 1976 William Collins Sons and Co. Ltd. U.S. edition: E.P. Dutton 1977.

Winged Pets: The Lure of the Falcon, pp. 33-34.
Reprinted by permission from *The Lure of the Falcon*, by Gerald Summers. Copyright © 1972 William Collins Sons and Co. Ltd. Fontana 1975. U.S. edition: Simon and Schuster.

Hats Off to Cats, pp. 38-39.
Reprinted by permission from *The Animals Came in One by One*, by Buster Lloyd-Jones. Copyright © 1966 Secker and Warburg Ltd. Fontana 1968.

That Mad, Bad Badger, pp. 43-44.
Reprinted by permission from *The Year of the Badger*, by Molly Burkett. Copyright © 1972 Macmillan, London and Basingstoke. Tandem 1975. U.S. edition: J.B. Lippincott Co. 1974.

The Komodo Dragon, pp. 51-53.
Reprinted by permission from *Zoo Quest for a Dragon*, by David Attenborough. Copyright © 1957 Lutterworth Press.

The Pride of Lions, pp. 54-56.
From *Born Free*, by Joy Adamson. Copyright © 1960 by Joy Adamson. William Collins Sons and Co. Ltd. Harvill Press. U.S. edition: Pantheon Books.
and from *Living Free*, by Joy Adamson. Copyright © 1961 by Joy Adamson. William Collins Sons and Co. Ltd. Harvill Press. U.S. edition: Harcourt Brace Jovanovich.

A Cheetah's Honor, pp. 57-58.
From *The Searching Spirit*, by Joy Adamson. Copyright © 1978 by Joy Adamson. William Collins Sons and Co. Ltd. Harvill Press. U.S. edition: Harcourt Brace Jovanovich.

A Beaver's Home, pp. 59-63.
From *Centennial*, by James A. Michener. Copyright © 1974 by Marjay Productions, Inc. Reprinted by permission of Random House, Inc. U.K. edition: Secker and Warburg 1974.

Wild Dogs: Solo in Danger, pp. 69-70.
Reprinted by permission from *Solo: the Story of an African Wild Dog*, by Hugo van Lawick. Copyright © 1973 by Hugo and Jane van Lawick-Goodall. William Collins Sons and Co. Ltd. U.S. edition: Houghton Mifflin Co. 1974.

The Death of a Bird, p. 78.
Reprinted from *Birds in Town and Village*, by W.H. Hudson. 1923. AMS Press 1968.

Monster Spiders, pp. 93-94.
Reprinted from *The Naturalist on the River Amazons*, by Henry W. Bates. 1863. E.P. Dutton 1969.

More Jaws, pp. 106-109.
Excerpts from *The Shark*, by Jacques-Yves Cousteau and Philippe Cousteau. Copyright © 1970 by Jacques-Yves Cousteau. Reprinted by permission of Doubleday and Co., Inc. U.K. edition: Cassell Ltd.

My Wild Life, pp. 117-120.
Reprinted by permission from *My Wild Life*, by Jimmy Chipperfield. Copyright © 1975 Macmillan, London and Basingstoke. U.S. edition: G. P. Putnam's Sons 1976.

The Big Cats, pp. 121-123.
From *Facing the Big Cats, My World of Lions and Tigers*, by Clyde Beatty. Copyright © 1964, 1965 by Clyde Beatty and Edward Anthony. Reprinted by permission of Doubleday and Co., Inc. U.K. edition: William Heinemann Ltd.

A Circus Year, p. 124.
Reprinted by permission from *A Circus Year*, by Michael Marden. Copyright © 1961 G.P. Putnam's Sons.

The Old Showmen Knew a Few Tricks, pp. 125-127.
Reprinted from *Seventy Years a Showman*, by "Lord" George Sanger. 1910.

Zoo Babies, pp. 128-130.
Reprinted by permission from *Cubs, Calves and Kangaroos*, by Paul Steinemann. Copyright © 1963 Orell Füssli Verlag. Elek Books 1965.

From Circus to Game Park, pp. 132-134.
Reprinted by permission from *My Wild Life*, by Jimmy Chipperfield. Copyright © 1975 Macmillan, London and Basingstoke. U.S. edition: G.P. Putnam's Sons 1976.

Temba and Tombi: A New Breed of Lion?, pp. 135-138.
Reprinted from *The White Lions of Timbavati*, by Chris McBride. Copyright © 1977 by Timbavati Private Nature Reserve. Paddington Press, Ltd. U.S. edition: Grosset and Dunlap. Bantam Books.

The Zoo, pp. 139-141.
Reprinted by permission from *Animal Magic*, by Johnny Morris, ed. by Douglas Thomas; by arrangement with the B.B.C. Copyright © 1966 David and Charles Ltd. U.S. edition: Taplinger Publishing Co., Inc.

News of Gnus, pp. 142-144.
Reprinted by permission from *Beasts in My Belfry*, by Gerald Durrell. Copyright © 1973 William Collins Sons and Co. Ltd. Fontana 1976. U.S. edition: *A Bevy of Beasts*. Simon and Schuster.

Dogs and Man: the Covenant, pp. 149-151.
Reprinted by permission from *King Solomon's Ring*, by Konrad Lorenz. Copyright © 1952 T.Y. Crowell. U.K. edition: Methuen and Co. Ltd.

Chimpanzee Society, pp. 152-154.
Reprinted by permission from *In the Shadow of Man*, by Jane Goodall. Copyright © 1971 William Collins Sons and Co. Ltd. Fontana 1974. U.S. edition: Houghton Mifflin Co. 1971.

Man and Baboon, pp. 155-157.
Reprinted by permission from *The Human Zoo*, by Desmond Morris. Copyright © 1969 by Desmond Morris. McGraw-Hill Book Co. U.K. edition: Jonathan Cape Ltd.

Animals Can Be Almost Human, pp. 158-159.
Reprinted by permission from *Animals Can Be Almost Human*, by Max Eastman. Copyright © 1957 Saturday Review Associates, Inc.

Pigeons and People, pp. 162-164.
Reprinted by permission from *Animal Magic*, by Johnny Morris, ed. by Douglas Thomas; by arrangement with the B.B.C. Copyright © 1966. Extract by Tony Soper. Copyright © 1966 by Tony Soper. David and Charles Ltd. U.S. edition: Taplinger Publishing Co., Inc.

The Camel—Discontented Ship of the Desert, pp. 165-166
Reprinted by permission from *Laura Was My Camel*, by Arthur Weigall. Copyright © 1962 J. B. Lippincott. Co.

Adaptation Techniques: Bat versus Moth, pp. 169-171.
Reprinted by permission from *The Magic of the Senses*, by Vitus Droscher. Copyright © 1966 W.H. Allen and Co., Ltd.

Horse Sense?, pp. 178-179.
Reprinted by permission from *Talking with Horses*, by Henry Blake. Copyright © 1975 Souvenir Press Ltd. Hodder and Stoughton. U.S. edition: E. P. Dutton 1976.

Horses and Telepathy, p. 208.
Reprinted by permission from *Talking with Horses*, by Henry Blake. Copyright © 1975 Souvenir Press Ltd. Hodder and Stoughton. U.S. edition: E. P. Dutton. 1976.

The Great Flamingo Rescue, pp. 216-219.
Reprinted by permission from *The Amazing World of Nature*, by Katharine Drake. Copyright © 1969 Reader's Digest Association.